WINDMILLS AND
MILLWRIGHTING

STANLEY FREESE

WINDMILLS AND MILLWRIGHTING

South Brunswick and New York

GREAT ALBION BOOKS

© Stanley Freese 1957, 1971

First American edition published 1972 by
Great Albion Books, a division of
Pierce Book Company, Inc.,
Cranbury, New Jersey 08512

Library of Congress Catalogue Card Number: 78-38803

ISBN 0-8453-1147-6

Printed in Great Britain

CONTENTS

LIST OF PLATES
(FOLLOWING PAGE 168)

1. Friston Mill, Saxmundham, in full sail, Easter 1934. (To be preserved.)

2. Bledlow Ridge Mill, High Wycombe, which had three cross-trees and six quarter-bars, but was very old and was pulled down in 1933. This and Cuddington Mill both had a circle of castors beneath the body. (Photographed in 1904.)

3. Stevington Mill, Bedford, 'discovered' by the author in 1929, is now preserved. It has been reconstructed and has clockwise sails. (Photographed in 1938.)

4. Cuddington Mill Aylesbury. The best post-mill in Buckinghamshire, seen in full sail in 1894. Demolished about 1925. Reefing ropes are visible on the sails; also the brake-rope trailing over the round-house roof.

5. Monk Soham Oak Mill, Suffolk, had the body supported on the round-house wall but no main-post. The sail-stocks were mortised through a wooden wind-shaft.

6. Shipley Mill, Sussex, in good order but disused, 1935. Built in 1879 for £2500. Note the grapnel irons behind the sails.

7. West Wratting Mill, Cambridgeshire, built in 1726. Note the two patent and two cloth sails, and the tail-pole. (Photographed in 1936.)

8. Cranbrook Mill, Kent, 72 feet high. Patent sails and fantail. Built by Humphreys in 1814 for £3500. Maintained by Messrs J. Russell & Sons.

9. Old Marsh Mill on the River Bure at Tunstall Bridge, Norfolk, with tail-pole and common sails with cloths furled, and paddle-wheel. Now gone. (Photographed in 1937.)

10. A Sussex landmark at Polegate in 1935. Note the five-bladed fantail. The sail alignment shows that one stock is weak and ought to have clamps fitted.

11. Heckington Mill, Lincolnshire, rebuilt in 1892, the last eight-sailer in England. The machinery came from a mill at Boston Docks, and was installed by the miller in a burnt-out tower. The sails have ¾-in. tie rods. Now preserved. (Photographed in 1934.)

12. Substructure of a post-mill. (a) Cross-trees and quarter-bar at Great Hormead Mill, Hertfordshire, resting on wooden plates. Note double abutment and tenon beneath iron strap, and main-post centred over cross-trees. (b) Iron-bound main-post of Pettaugh Mill, near Stowmarket, showing 'wood wears' in orifice on floor, and upright stay-rod.

ix

13. The framing of windmills. (*a*) The body of Drinkstone Mill, Bury St Edmunds, was reversed back to front many years ago. Built in 1689 the mill is now disused and has two cloth and two spring sails. (Photographed in 1938.) (*b*) Mr Denis Sanders's exact scale model of Shiremark smock-mill, Surrey, with cap frame in place. (*c*) How not to build a post-mill: Upper Arncot Mill, Oxfordshire, in 1945. No diagonal braces or tie-rods; side-girts mortised to crown-tree; quarter-bars spliced. Mill stones are on a hurst frame on bottom floor. The mill has since collapsed. (*d*) Correctly designed mill at Tunstall, Suffolk, with braces and tie-rods; being dismantled owing to lack of trade.

14. Wind-shafts and bearings. (*a*) Rear view of neck-brass and pot on neck-beam from Rock Hill Mill, Sussex. The brass has lips to prevent it from revolving or moving forwards, and an oil trough at top edge. Note tail brass at right-hand side. (*b*) Filleted iron neck-wear in wooden shaft at Rougham Common Mill, Bury St Edmunds, believed built in 1770. (*c*) Canister at Haughley Mill, Suffolk (post-mill, burnt down), showing fins and bolts and retaining ring. (*d*) Mortised wooden wind-shaft head of Finchingfield Mill, Essex, built about 1760. (*e*) Square shaft with mortised canister, at Arncot Mill, Oxfordshire. Iron band brake lies across neck.

15. Stone-floor and ground-floor. (*a*) Stone floor at Westleton Mill, Suffolk, with great spur-wheel above 'wallower'. Part of wooden brake is visible; also auxiliary take-off, on left, and stone-nut on crutch-pole over millstone. (*b*) Lower sprattle-beam of East Wittering Mill, Sussex (ground floor), wedged at end, with foot of main upright shaft; also bridging-box containing toe-brass. Wedged bridge-tree on left, with mortised brayer in background.

16. Wheels, millstones and governors. (*a*) Great spur-wheel of Stelling Minnis Mill, Kent, with auxiliary drive beneath; and stone-nut, jack-ring and bridge-tree on left. Derbyshire Peak bed-stone below. (Photograph by Mr Denis Sanders.) (*b*) Feed-shoe with damsel (beneath the hopper) and spring rod and adjusting cord on stone-case, at Lower Woolstone watermill, Buckinghamshire. (*c*) French burr runner-stone. (*d*) Lag governor at Normanton Common post-mill, Derbyshire. (Photograph by Mr Paul Baker of Nottingham.) (*e*) Centrifugal governors, and stone-nut with wrought iron quant or crutch-pole in Lacey Green smock-mill, Buckinghamshire.

17. Bolter and dresser, Upminster smock-mill, Essex. (*a*) Bolter casing and reel. (*b*) Old flour dresser, belt-driven. (Mill now preserved.)

18. Moving a main-post. Weston endless chain double-pulley tackle. Note how slack portion of chain should be draped over taut portion to make it feed smoothly. The main-post is being removed from a disused round-house.

19. Main-post and weather-beam. (*a*) Head of main-post at Ashfield, Suffolk, showing collar of 'Samson head' and the wooden pintle. (*b*) Prick-post, weather-beam, and diagonal braces in a well-designed Suffolk post-mill, with stay-rods and weather-stud irons above them; upright shaft, driving gears and adjustable spindle beam.

20. Tackle and tools. (*a*) Moving a wind-shaft with Weston endless chain gear. The canister is held by rope-tackle and a sling-chain suspended from a sail-stock lashed to the mill. Note 'cheek' bolted to weather-stud (centre of

picture) and strut supporting the latter. (*b*) Lifting post-mill steps with a five-ton jack (inset) to fit tram wheels, etc. (*c*) The actual bull-nosed auger used for drilling Thorpeness Mill main-post; also an old shell-auger and a screw-auger.

21. Mill carpentry. (*a*) Clasp-arm brake-wheel torn out of Wenhaston Mill, Halesworth, in a gale in 1946. (*b*) The author fixing a dummy sail-cloth on Brill windmill. (Photograph by Mr L. J. Reason.) (*c*) The date of Pitstone Mill, Buckinghamshire, the oldest in England—1627. Recently discovered by the author, and here retouched to bring it into prominence.

22. Cogging the mill. (*a*) After sawing out in 'blank' form, the centres of all the cogs are found and the outlines very accurately marked off with a special gauge. (*b*) Mortised iron wheel with wedged cogs at Upminster. Gimbals, stone-spindle and footstep in centre; and mace, bridge-piece, and 'damsel', mounted on another stone-spindle, on right.

23. Spreading a cloth at Brill windmill. (*a*) Sail-cloth at 'sword-point' on one of the new (1948) sails. Note bowsprit and strainer wires on sails. (*b*) Spreading out the cloth to 'dagger-point' with the pointing ropes, which will be drawn round the back of the sail-frame and made secure.

24. Spring sails. (*a*) Detail of spring sails used in conjunction with common sails, at Earnley Mill, Chichester. The control rod from the spring is hooked on to the sail-rod. (*b*) Icklesham Mill, Sussex, in 1936, with full elliptic spring sails; short clamps on stocks; fantail on roof, driving down to the step wheels; and part wooden round-house. (*c*) Wrawby Mill, Lincolnshire, with spring sails on an iron cross ; and tail pole. Since completely renewed by John Sass and Edward Travis. (Photograph by Mr H. E. S. Simmons.)

25. Patent sails. Close-up of patent sail at Wenhaston Mill, Suffolk, about 27 ft. × 8 ft., depicting 'spider' and triangles and sail-rod controls.

26. Two old millers. (*a*) The late Mr Steve Aldred in 1947, reputed to have been the best wheat flour miller in East Anglia. His mill at Saxtead Green was always swept clean, and is preserved. (*b*) The late Mr J. B. Geater at the age of 93, with a picture of his old mill at Stratford St Andrew, Suffolk.

27. Fantails and staging. (*a*) The 77-ft. Haverhill Mill, Suffolk, with 50 ft. wind-wheel, had a special outrigger stage for reaching the wheel. (Photographed in 1938.) (*b*) View of New Bolingbroke Mill in 1935, showing typical Lincolnshire assembly of fantail and fanstage with lever-type reefing gear, and ogee cap. (Photograph by Mr H. E. S. Simmons.)

28. Stone-dressing. (*a*) Mr John Bryant of Pakenham lifting a runner-stone with a worm-tackle and 'clam' in the windmill that regularly appears on television. (*b*) Mr J. Wightman, expert flour windmiller, illustrates the dressing of an anti-clockwise stone. (*c*) Mr H. Barker of Norton, Suffolk, dresses a clockwise stone at Pakenham. In foreground is a tracer-bar for checking the stone-spindle; and a 'clam' for lifting millstones; also a number of mill-bills.

29. New sails at Brill. (*a*) The last sail and stock are down, and two old stagers ask 'what next?' (*b*) Mr Jack Blake assembles new sail frames in Mr Green's yard; but will they get as far as the mill? (*c*) A new stock is wheeled across the common; the quarter-bar is strengthened in readiness. (*d*) Head of mill supported; stocks in place; the first sail goes up quickly and smoothly, 3 October 1948.

30. A common-sailed paddle-wheel mill. This well-kept mill overlooks the Waveney, on Herringfleet marshes, and is preserved by Lord Somerleyton.

31. Sail-repairing. (*a*) Sail frame laid out on stock at Pakenham for checking bolt holes. Note 'angle of weather'. (*b*) Neck brass and 'swing-pot' removed. Note shoulder and extended curve at leading side of brass (nearest the camera) to receive the thrust as the shaft revolves.

32. The late Miller Greenard being photographed by Mr H. E. S. Simmons in 1934, in front of the beautifully kept Worlingworth Mill, Suffolk, now demolished.

33. Cross-in-Hand Mill, largest post-mill in Sussex, with fantail on tail-pole, and wooden-walled round-house. The model in the foreground was made by Syd Ashdown, the miller's son. Now preserved. (Photographed in 1935.)

34. Headcorn Mill, Kent, caught on a lovely day in 1936, with typical broad stage supported from the ground. Note wheel-and-chain emergency winding gear, and lever-type reefing gear. (Mill now completely gone.)

35. The modern Coleby Heath six-sailer, Lincolnshire, destroyed by the Air Ministry to safeguard low-flying aircraft. (Photograph by Mr H. E. S. Simmons.)

PREFACE

IN England less than a century ago the windmill was a sight so common that it went almost unregarded, except when some writer or some painter bestowed particular attention upon it.

Artists have always had a fondness for windmills. There are famous windmill paintings, including one that is reputed to have been sold for £10,000 to an American before the war, at a time when half-a-crown could only with difficulty be raised for the proposed preservation of a real working windmill in the country. A century-old painting, John Crome's picture of Mousehold Heath Mill, Norwich, hangs honoured at the National Gallery in London. A son of the last miller of this very windmill has written about it, and one of the finest windmill models in England—the late H. O. Clark's model in the South Kensington Science Museum—portrays this same windmill in an exact scale-reproduction that occupied its modeller's spare time for fifteen years.

In 1916 the late Allen Clarke brought windmills to life again in an account of the windmills of the Fylde of Lancashire; and in the last twenty years several books have been devoted to the windmill, which (in common with other institutions of more leisurely times) comes in for increasing appreciation now in what is the twilight of its existence.

My own enthusiasm began with a holiday spent at Darsham in Suffolk, in 1923, when I had already acquired a casual interest in windmills.

The windmill, you may say, is a familiar landmark in Holland too; well, so it is; but many Dutch windmills are nothing but water pumps; they lack the wonderful old corn-milling machinery, without which, to my mind, a windmill is only half-a-windmill. So I propose to write here only of English windmills; and in setting forth for Darsham so long ago, I believe I was, though unwittingly then, aiming at about as good an English windmill as ever was built.

On my way to Darsham, I specially remember stopping to admire the white-boarded post-mill (now demolished) at Leavenheath (Sudbury), its paintwork shining in the fierce sunlight of a heat-wave—a post-mill is one with a rectangular wooden-boarded body, the whole turning round a central post or tree, to face the wind. Later in the day I passed the old tumble-down Kelsale post-mill on the Lowestoft road, which soon went the way of so many others; and I remember hoping, as the dusty road (as yet untarred) rolled up and down over undulating country, that the cottage guest-house at Darsham where I was to stay would be on high ground—I preferred hilltops to valleys—and that it would be near to a certain windmill marked on my Ordnance map.

My wishes were realized, for this lovely, well-kept, white-boarded mill, surrounded by acres of golden corn, which it would presently grind, stood right opposite the cottage, to be seen from dining room and bedroom, and I was surprised when I woke up in the morning to find the mill facing the other way—I had not realized then that the whole mill would turn itself round as often as the wind changed, even while the miller slept!

For the rest of the holiday I followed the fortunes of this windmill, my neighbour, whose sails turned for one or two of the days, despite the sultry summer weather. And the adjoining countryside, I noted, was made lively by the white sails of many other windmills turning in mill-yards or across the broad commons of gorse and heather for which east Suffolk is famous.

William Cobbett, over a century ago, when some ten thousand working windmills enlivened the English country scene, was struck by the beauty of the white East Anglian post-mills, although he could, of course, have discovered similarly enlivened landscapes in his own favourite Sussex. Of the scene in and around Ipswich he wrote: 'The windmills on the hills in the vicinage are so numerous that I counted, while standing in one place, no less than seventeen. They are all painted or washed white; the sails are black; it was a fine morning, the wind was brisk, and their twirling altogether added greatly to the beauty of the scene, which, having the broad and beautiful arm of the sea on the one hand, and the fields and meadows studded with farmhouses, on the other, appeared to me the most beautiful sight of the kind that I had ever beheld.'

I made a little colour sketch of Darsham Mill in 1923, and later began to record windmills more extensively by drawings done with a fountain pen: and I learnt much about them by sketching interior details and sectional elevations and plans, as well as general external aspects.

Then I turned to photography, and gradually got together hundreds of pictures of interiors and exteriors, some of which have been chosen to illustrate this book.

But neither photograph nor drawing can capture for my pages the most beautiful of all country scenes; only those who have caught sight of the big white sails of a windmill following one another against a background of dark hills or woodlands, or black canvas sails soaring one after another into an evening sky, can fully apprehend the characteristic beauty of this structure, which differs from all others because it is alive with comely motion, never awkward or ungainly, but blending well with every kind of landscape.

So I, having been gripped by this passion for windmills, set out some twenty-seven years ago with sketch books and pen on a tour of the eastern counties, determined to have a go at the windmills of England. On this excursion, by the hospitality of the windmilling fraternity, I was invited into a number of these windmills, where I stopped to sketch them; and one I particularly remember and am very glad to have been in—it has since been blown down—was old Mr Stephenson's open post-mill at Friskney Tofts in Lincolnshire —as fine an example as could be found of the earlier type of post and tail-pole mill with cloth sails.[1]

How proudly the old miller pointed out to me the excellent workmanship and the ingenuity of the structure and machinery of his old mill; and who could fail to be agog, as I certainly was, to inspect the interiors of many more windmills up and down the country? I have in fact since visited, and noted down in diagrammatic and perspective

---

[1] An open post-mill is one without a brick-built enclosure to protect the main centre-post where it is exposed to the elements, beneath the body; such mills were usually pushed round to the wind with a long 'tail-pole'.

outlines, the interior economy of more than a hundred windmills from Sussex and Kent to East Anglia, the Midlands and Dorset and Carlisle, and enjoyed every minute of the time. Meanwhile, in conversation with the many millers I met, I sought to find out in more and more detail exactly how these wonderful structures were fashioned and erected by the country millwrights.

Then one day I found one of these masters of the millwrighting profession—a septuagenarian like most of his surviving colleagues —engaged upon an important windmilling job, and he lavishly and willingly added to the knowledge already imparted to me by friendly millers.

Today, when the days of these craftsmen, and of the windmills themselves, are numbered, there is still work to be done in the preservation of our remaining windmills, at least for instruction and education, although the large majority have, alas, ceased to grind corn. But for such work we need *working parties* of practical enthusiasts, who can even saw a timber, handle a rope-tackle, or work from a suspended 'cradle'; we do not want *windmill committees* of the kind who will argue and obstruct and interfere with progress, and pop up at the end to get the major credit for other folk's work.

Millwrights are passing away one by one, leaving behind them products of a craftsmanship that cannot now be matched. Let us remember, then, that a millwright is one who constructs mills or their machinery—a wright is a workman, especially in wood—an artificer, an individual craftsman whose workmanship and skill helped to make England what she was when the windmill and the watermill ground and dressed our flour (with here and there solidly built steam-engines standing nearby ready for days of drought or becalmed windmill sails). The millwright is mortal, alas, and disappearing; but the windmills he built so well will stand as his monument for some years yet, to remind us all of the days when men built to last, for the windmill was essentially a long-term investment.

This book is an attempt to record briefly on paper the work of those craftsmen of the past who surrounded themselves with the special tools and tackle used in the making of a windmill, with their little forge and their bricklayer's trowel, their improvised lathe and small iron foundry, all of which they knew how to use with nothing less than mastery.

For the information contained in these pages I am indebted to millers too numerous to mention, and to several millwrights and mill repairers—Mr Jesse Wightman, whose fund of knowledge covers both practical milling and millwrighting; the late Mr Hector Stone, the octagenarian miller of Upminster; Mr Ted Friend of Suffolk; Mr Walter Rose of Haddenham, Buckinghamshire; the late Mr A. Clarke, willing to impart his endless store of information about mills (some of the best ones he demolished on government orders during an invasion panic in the 1914 war); and the late Mr Will Cripps of Cuddington, Buckinghamshire—all descended from milling and millwrighting families; and of course I have learned much from Mr Rex Wailes's writings on windmills.

To Mr Syd Simmons of Shoreham, Mr Denis Sanders of Feltham, and Mr Paul Baker of Nottingham, I am indebted for permission to use some photographs acknowledged to them in the captions. The others were taken with my own camera.

A man to whom special acknowledgement is overdue is the late Mr H. J. Smith of Mears Ashby, Northants, whose letters to *Cycling* in 1927 and 1928 laid the foundations of the movement to record and photograph all the windmills of England

May the sails continue to turn upon at least a few surviving mills for many a generation to come; and may the Ministry of Town and Country Planning, if it has the power, do as the Secretary-General of the Department of Education, Art and Science is said to have done in Holland (by order dated September 1942)—forbid the demolition of any mill, or alteration of its use, unless wholly made of metal (i.e. the ugly and inefficient wind-pump) without his consent; and may this power come and be used before it is too late.

S. F.

September 1956

ENCOURAGED by the continued demand for my book—at Windmill Lectures, and whilst showing visitors over the various windmills I have personally helped to restore— I have now revised and corrected the text so far as is practicable under the present process of reproduction.

It is hoped that the windmills I and my friends have worked upon will thus be indefinitely preserved. Particularly at Pitstone, the work of the Pitstone Windmill Preservation Committe has been very successful where a seventeenth-century windmill, rebuilt from an older one in 1624 and modified more recently, has been restored to working order, though only to grind occasionally under strict supervision. This earned the windmill a 'Countryside in 1970' award and plaque under the Conservation of the Countryside Competitions in that year.

Other windmills have been voluntarily restored or partly so, such as North Leverton, Willingham, Over, Compton Wynyatts, Chesterton and Wrawby in or near the Midlands; White Mill, Sandwich (Kent), Holton and Framsden (Suffolk), etc, by Graham Wilson, Derek Ogden (now a professional), Vincent Pargeter (now a professional with Philip Barrett-Lennard), John Sass and Edward Travis in Lincs, Chris Hullcoop and Brian Flint with Marcus Cook and Adrian Coleman in Suffolk; also Jeff Hawkins, David Wray, Tom Owen and the present writer; and a host of other willing hands, at Framsden and Pitstone especially.

Thus has the total number of preserved windmills in England increased from about ten to twenty, I am told, in the last ten years. And the number could continue to increase if those enthusiasts who are able, will turn their hands to the enthralling and satisfying pastime of windmill repairing whenever they have an hour or two to spare, a deserving windmill in the neighbourhood, and some support from the Society for the Protection of Ancient Buildings at 55 Great Ormond Street, WC1, and other bodies.

S.F.

October 1970.

Chapter I

CONCERNING WINDMILL HISTORY
AND INVENTIONS

❧

WINDMILLS, if Sir Richard Phillips was right (*Million of Facts*, 1841, p. 1027), were invented by Hero in the first century B.C. (this means, of course, Hero of Alexandria, whose scientific writings are well known); but the claim is vague, and proof would be elusive. The late Allen Clarke said he was told that in the *Arthasastra* of Kautilya, an old Hindu classic dated about 400 B.C., water is said to have been raised 'by contrivances worked by wind-power' (*More Windmill Land*, p. 565).[1]

ORIGIN OF WINDMILLS

It is generally supposed that windmills originated in the East, and that the Third Crusade (1189–92) was responsible for their introduction to western Europe. Beckmann, however, in his *History of Ancient Institutions, Inventions and Discoveries* (1823 ed., Vol. II), asserted that windmills existed in Europe during the First Crusade, 1096–9, and were common in the twelfth century, and he questioned whether there were at this time, or at any other, any considerable number of windmills in the East.

It is difficult to establish the existence of any windmill in England before the time of the Third Crusade. Medieval references to windmills, although rare at any earlier date, become frequent from the close of the twelfth century.

Domesday Book records many 'mills', but does not specifically mention a windmill; and the generally accepted inference from this is that all the Domesday mills were either water-driven, or, where there

---

[1] Quoted to Mr Clarke from *Studies in Ancient Hindu Polity*, by Narenda Nath.

was insufficient water, cow-mills or horse-mills of the kind fully dealt with in Bennett and Elton's *History of Corn Milling* (1898)—a standard work, though scarce, embracing mortars, hand-querns, treadmills, horse- and cow-mills, watermills and windmills, and of interest to every technically minded student of windmills.

Sharon Turner, the first historian to explore the Anglo-Saxon manuscripts of the British Museum, says in his *History of the Anglo-Saxons*, printed in 1805 (Vol. IV, p. 62): 'Their corn was thrashed with a flail, like our own, and ground by a simple mechanism of mills, of which great numbers are particularized in Domesday Survey. In their most ancient law we read of a king's grinding-servant; but both windmills and watermills occur very frequently in their conveyances after that date.'

Turner, who is reputed to have been cautious in his pronouncements, does not seem to have answered a request in the *Gentleman's Magazine* of December 1805, that he should produce his authorities. It seems unlikely that the true origin of the windmill will ever be discovered. Indeed, it may have been invented in more than one country quite independently.

The windmill was not a common feature of the English landscape until about a century after the completion of the Domesday Book; and Fairbairn points out in *Mills and Millwork* (4th ed., 1878, p. 272) that Bishop Puiset's *Boldon Buke* of 1183 mentions only horse-mills and watermills.

The first generally accepted authentic instance of an English windmill occurs in *Jocelyn's Chronicle*,[1] which refers to a windmill that Dean Herbert erected upon his glebe lands sometime between 1182 and 1203.

Other documentary references to windmills occur in 1185 at Weedley, Yorkshire, and in 1165 during the building of Orford Castle, Suffolk, in that year.

A twelfth-century windmill is also mentioned in the *Cartulary of Oseney Abbey* (Oxford Historical Society, 1935), edited by H. E. Salter (Vol. V). In Charter no. 692, Henry Doily, Constable of King

---

[1] Jocelyn de Brakelond's *Chronicon*, Ch. 43, translation in Thomas Carlyle's *Past and Present* (1843), Book II, Ch. 15.

Henry II, grants to the Abbey free entrance and exit across his land in 'Cleydone', to the Abbey's windmill in 'Cleydon' (Steeple Claydon, near Buckingham). The charter is undated, but the reference to King Henry II places it between 1154 and 1189.

WINDMILL TYPES

For lack of records, too, the date of the invention of the three usual classes of windmill (post-mills, smock-mills and tower-mills) is likely to be always a matter of conjecture; the best one can do is to point to the earliest recorded instance of each type.

Most medieval representations of windmills show a revolving post-mill; out of eight examples of medieval windmill sites known to John Salmon ('The Windmill in English Medieval Art', in *Journal of the British Architectural Association*, 1941) *all* bore traces of cross-trees buried in the ground, proving that they were post-mills of the 'sunk' or 'peg' type. (Mill mounds surviving from sunk-post-mills look rather like large hot-cross buns, because the horizontal cross-timbers which centred the upright main-post were located underground.) Perhaps fixed windmills existed in England at the same time as post-mills in the sixteenth century (see the accompanying reproduction (Fig. 1) from the 1652 reprint of *England's Improvement, or Reducement of Land to Pristine Fertility*, by Walter Blith). In passing, we may note that the cloths on the type of sail depicted here were usually interlaced through alternate sail-bars, according to pictures of early Continental mills; this may have been a survival of the practice of threading rushes or reeds through the bars, but purchase of cloth for sails and cords for reefing them is recorded back to 1300.

We might visualize an early fixed mill, roughly shaped with wood-man's tools, with, carrying the sails, a horizontal wind-shaft geared to a single small pair of millstones—little more than a square-built shed with sails spread with coarse cloth, or even intertwined with long rushes.

There is a tradition among old millers that the earliest windmills had no proper brake, though the present type of brake-assembly was in use in the sixteenth century.

The Mill made open that the whole Engin may appeare

Fig. 1. An early fixed windmill (from the 1652 reprint of Walter Blith's *England's Improvement*).

POST-MILLS

Our first windmills were probably sunk-post-mills, that is to say, a mill with its post deeply embedded in the ground like a gate-post; and a main-post apparently of this type is seen in Allen Clarke's photograph of Hambleton Mill post in his *Windmill Land* (Vol. I, p. 262). Generally the sunk-post-mills did not greatly differ outwardly from the more familiar open post-mills, except that the body was usually close to the ground. Originally the body was just a square single-floored affair with a pent-house roof (the 'ogee' roof coming later), and in some early illustrations there are no 'quarter-bars' or supporting props to the main-post. By the end of the Middle Ages, the 'post-and-trestle' mill, with a tail-pole for pushing the mill round into the wind, and 'common' or cloth sails, had probably become almost standard; it was built in greater numbers in England than any other kind.

SMOCK-MILLS AND TOWER-MILLS

But long before the post-and-trestle mill ceased to be built, an alternative design had been evolved—the 'smock-mill', of which Lacey Green Mill, Buckinghamshire, originally erected at Chesham in 1650, seems to be the first recorded example. The smock-mill, or frock-mill, is a wooden-walled building, usually octagonal, with 'battered' or sloping sides so that the sails can pass round it without having the body too small or cramped at the base, nor too large and clumsy at the top; the body of the smock-mill is fixed, only the cap and sails are free to turn about to face the wind.

Brick- or stone-built 'tower-mills' are thought to have come into general use in this country at about the same period (mid-seventeenth century), though a fifteenth-century print of Rhodes Harbour, in the Aegean Sea, shows a row of stone-built tower-mills on the jetty (Breyerbach's *Itinerarium*, 1486). These may have been 'fixed' mills with the sails always pointing into the prevailing wind, for John Adrian Leegwater, an inventor who died in 1650, was quoted by J. Beckmann (*History of Ancient Institutions, Inventions, etc.*) as saying that the mode of building a windmill with a movable roof was first found out by a Fleming in the middle of the sixteenth century.

Marsh-mills for draining the fenlands (of which Fig. 1 depicts a very ancient example) were usually of the smock or tower type.

HYBRID MILL TYPES

Two other types of windmill to be found in England may be mentioned, the hollow post-mill and the composite; the only example of the hollow post-mill is on Wimbledon Common, near London. This is constructed like a miniature smock-mill raised upon a big octagonal house, which contains the machinery. An upright driving shaft passes down through the fixed or smock portion from a brake-wheel contained in a cap shaped like a little post-mill body; and the sails are short enough to clear the roof of the main brick-built structure. It is difficult to see the advantage of this sort of mill; the sails are necessarily very small on English examples.

Secondly, there is the composite mill, which is usually an old post-mill reconstructed with the post removed, and the body carried not on a post but on castors or tram-wheels that run upon the wall of a strongly built circular shelter or round-house. For turning the mill and its sails into the wind, there is usually an automatic fantail on the *roof* of a composite mill, driving down to a cog-ring on the round-house curb, so that the structure serves altogether the same purpose as that of a tower-mill (on which the cap and sails are usually orientated by a fantail and gearing), except that it is carrying the whole mill body round instead of the cap only. Such a mill should not be confused with several true post-mills having a roof-top fantail coupled to a cog-ring on the round-house, as, for example, the old wreck (it collapsed in 1960) near Hargrave, on the Bedfordshire–Huntingdonshire border. Most fantail post-mills have the fantail mounted on a big frame above the steps which lead up to the mill door (see Pl. 1). At least one composite mill (at Monk Soham Oak, Suffolk, Pl. 5) was actually designed and constructed as a composite, never having had a main-post, though it resembled very closely indeed a typical Suffolk post-mill. Even the fantail or 'fly' was set

6

above the steps (the wheels of which it drove more or less conventionally), whilst the round-house appeared to have its own roof, whereas most composite mills have a clumsy overhanging 'skirt' extended out from the mill body to span the round-house wall, like the Midland post-mills. This mill, which may have been unique, was not very old, but was fraily constructed and has now been demolished.

Another little-known type of windmill, no longer extant in England, was the horizontal mill, of which type one was erected by Captain Stephen Hooper at Battersea. It was shaped like a tall gas-holder; it had the sails or vanes revolving on a vertical shaft, with an external casing of adjustable louvres by means of which the wind could be directed upon one half of the vanes at a time.

SAIL DEVELOPMENTS

The most primitive windmills discovered or recorded abroad were provided with a vertical rotating shaft, round which a series of upright evenly spaced vanes were attached by outriggers. These vanes were either placed radially as in Stephen Hooper's mill, with a shield around one side so that the wind drove only on the unshielded side and so rotated the sails, or they were set at a tangent rather like the modern chimney cowl.

But the earliest illustrated windmills in England were mills of the post-mill type, with the wind-shaft horizontal; in these the sails necessarily had to be made to revolve according to a different principle; and the unknown inventor who solved the problem established a basic principle without which (one might say) even propeller-driven aeroplanes themselves would probably never have flown. He discovered that if a set of, say, four vanes or sails were fixed on the end of a spindle and pointed into the wind, they would revolve if each individual vane were twisted a little edgeways out of the wind.

For several centuries this 'weather' or 'angle of weather' on the sails was made an angle constant from heel to tip, like a flat board turned at an angle. But in 1759 John Smeaton (1724–92) the millwright and lighthouse builder, in a paper entitled 'An Experimental

Inquiry concerning the Natural Powers of Water and Wind to turn Mills and other Machines depending on Circular Motion', read the results (to the Royal Society) of his carefully conducted experiments to determine the best design and dimensions for windmill sails. In this he formulated a table of angles for the 'weather' of the sail, which varied from approximately 18 degrees at the heel to 7 degrees at the tip, thus giving a sail the sort of twist that is seen in a propeller blade. All windmill sails are now shaped more or less in this way.

Smeaton also popularised the multi-sailed windmill—beginning with the five-sailed Flint Mill, Leeds, in 1774; but the Brazil combined smock-mill and watermill, which stood near Babers Bridge on Hounslow Heath as early as 1757, had six 'common' sails, according to a contemporary painting.

Another man of fertile brain, Andrew Meikle (the man who trained John Rennie), experimented with alternative types of sail. At one time cloth sails were universal—a single sheet of cloth spread by hand over each sail frame. In 1772 Meikle produced the 'spring' sail (Pl. 24) with a series of transverse rectangular wooden shutters controlled by springs and adjustable levers; and in 1780 he adapted a kind of centrifugal control for furling and unfurling a modified cloth sail having a series of short cloth rollers (Fairbairn, *Mills and Millwork*, p. 278). On this sail was a sliding frame, coupled to an adjustable weight by a long iron rod passing through the length of the wind-shaft. Increased velocity of the sails caused the centrifugal movement of the slide to furl the cloth; but the weight, counterbalancing the slides, unfurled the cloth again when the wind dropped and the speed fell. This device was not perpetuated, although it was mechanically satisfactory, perhaps because the rolling cloth shutters wore out too quickly; but Fairbairn states that it was the first successful automatic reefing apparatus to be applied to windmills. Meikle seems to have co-operated with Smeaton in perfecting this invention.

In 1807, William Cubitt, a millwright, coupled the rod-and-weight control mechanism to a set of pivoted rectangular wooden shutters as used on Meikle's spring sail; and this new modification, which became known as the 'patent' sail (Pl. 25), was more or less generally adopted for later mills throughout Britain, Germany and Denmark.

An annular sail with shutters controlled in the same way was patented in 1855, the year in which a well-known mill was built at Haverhill with a sail of this type.

An important innovation called 'Catchpole's Patent' was fitted to three or four windmills in Suffolk, including the exceptionally well-proportioned Buxhall tower-mill, which had been rebuilt from a smock-mill. This device consisted of a pair of longitudinal shutters fixed on the leading-edge of each sail at the tip, where, of course, they caught the wind best; when the vanes of the sails opened, these two shutters were also adjusted by the patent-sail controls so that they acted as a brake. It was simple and efficient, and it is surprising that such an excellent improvement seldom spread to other counties.

LUFFING GEAR

To luff a windmill is to turn it to face the wind. Primitive vertical-spindle windmills mentioned above required no 'luffing' or 'winding' gear, because the wind, from whichever quarter it blew, was able to play on one side or another of the sail assembly, but when the sails were attached to a horizontal wind-shaft—which proved on the whole to be more serviceable—means had to be devised for facing shaft and sails into the wind; otherwise the mill would only work in a prevailing wind (which it would have to be built to face); and in England, a country of changeable winds, this would be of little value.

A tail-pole was therefore attached to a post-mill (see Pl. 3), and by means of this long pole projecting at the tail of the mill, the whole structure could be turned round into the wind by the miller when necessary. Smock-mills and tower-mills had a similar pole suspended from the rear of the cap and strengthened with outriggers, but there were exceptions. Some tower-mills abroad, and possibly in England, had the cap levered round from within by hand by means of a big crowbar. An alternative method was by an internal capstan wheel that engaged with a cog-ring. But the tail-pole method eventually prevailed until some inventor unknown thought of gearing the cog-ring on the mill-curb to a pulley-rimmed wheel that could be turned from the ground by an endless chain over it, the chain hanging down from the tail or back of the cap so as to be within reach. Thus the miller could rotate the wheel, and luff the cap, from the ground. The internal capstan, which must in the same way have engaged with a cog-ring on the mill, was thus superseded. Hand-chain gear became common in mills in Anglesey and was very much used in the Fylde and Cheshire.

In 1745 a method was discovered for facing a windmill automatically into the wind; in that year Edmund Lee patented the fantail, widely adopted thereafter in England, Germany and Denmark. Andrew Meikle appears to have appropriated this idea to himself in 1750, though generally this great inventor, who also made the first successful threshing machine, was unduly modest and did not reap the full benefits of his own ingenuity (Smiles, *Life of John Rennie*, 1874 ed.). The fantail consists of a six- or eight-vaned wheel mounted behind the cap or the mill body in a plane at right angles to the plane of the main sails. The vanes of the fantail are inclined, so that the wind revolves it, and this through suitable gearing turns the cap round slowly to the left or right according to which way the wind veers. It has been pointed out that the fantail, with its train of small iron gears, only became practicable as the result of improvements in the technique of iron-founding.

In 1752 Mr Cowper, of Pennyfields, Poplar, erected on the top of Newgate Prison a windmill for driving a series of devices like bellows in the various cells, for ventilation; and the 'sails' of this contraption took the form of an eight-bladed fantail 14 ft. 6 in. in diameter, exactly conforming to the design and construction of a windmill fantail. This sail or fan was mounted upon a crankshaft which drove a rod up and down through a drilled oak main-post (provided with cross-trees and quarter-bars like a post-mill), and a vane projected at the tail to luff the fan, this vane being embellished with the coat of arms of the City of London. The main-post had no pintle; the weight was carried by a set of $5\frac{1}{2}$ in. castors running upon a girdle on top of the quarter-bars.

MILL GEARING

Any early English windmill with a horizontal wind-shaft must have been geared to the upright spindle of the millstone much as it is to this day. This was done by mounting a face-wheel on the wind-shaft, the cogs being merely short round staves projecting from the face of the wheel, near the periphery, these cogs engaging with a 'lantern pinion' on the stone-spindle. Lantern pinions consisted of two wooden disks spaced apart on the spindle by a series of staves parallel to the spindle (see Fig. 1), or the pinion might have only one disk, with staves projecting from it as in the face-wheel. This pinion would be known more particularly as a 'pin-gear' or 'trundle wheel', although the term was sometimes applied to lantern pinions too. Nowadays all cog-wheels working at right angles to each other have bevelled cogs, but this refinement seems to have been discovered only about two centuries ago. Before this, square-cut cogs began to supersede the lantern pinions or trundles. Wheels of this type, normally working together in the same plane, are called 'spur' wheels. In the interim between lantern pinions and bevelled cogs, a right-angled drive sometimes consisted of a spur pinion on the wallower wheel, engaging with a face-wheel with straight cogs on the wind-shaft (as still surviving at Lacey Green Mill, Buckinghamshire, which has the original gears of 1650, or a replica of them).

By this time it had become the practice to provide a pinion with one-half or one-third as many cogs as the driving wheel—the wheels at Lacey Green have 36 and 72 cogs—this would certainly simplify calculation of relative wheel speeds.[1] But later on an extra or 'hunting' cog was provided, giving, say, 36 and 73; the purpose being to prevent identical cogs from meeting at every second or third revolution, which would have eventually exaggerated uneven wear due to inequalities in the meeting cog faces.

In 1752, about the time that cast-iron wheels began to be manufactured, Camus worked out the effects of epicycloidal curves on the working faces of spur and bevel cogs to secure smoother meshing,

[1] An ancient little mill at Yoxford (Suffolk) had 72 cogs engaging with 18 staves in a lantern pinion.

and to reduce 'back-lash', and a little later Euler and Kaestner developed the 'involute' curvature which, however, found less favour than the epicycloidal shape for windmill gears (Fairbairn, *Mills and Millwork*, p. 292). Samuel Smiles stated that James Watt's assistant, William Murdock, cast the first iron-toothed gears in 1760.

Fairbairn also says, presumably on the authority of Smiles, that cast iron was introduced into mill-work by John Smeaton at the Carron Iron Mills near Falkirk in 1769, and more extensively by John Rennie (the builder of London Bridge, etc.) in 1785 at the Albion Steam Flour Mills, Blackfriars. Smeaton himself says that he introduced cast iron into mill-work at Wakefield windmill in 1755. Very soon shafts and wheels of iron came to be applied to windmills and watermills, but wood work was never entirely displaced by iron work even in the most modern examples of the windmill.

LAYOUT OF MILLSTONES: HEAD-AND-TAIL MILLS, ETC.

Many English windmills had only a single pair of millstones, one of which was fixed in the floor, while the other was made to revolve upon it by means of gearing; but when experience made it possible to build larger post-mills—perhaps 300 years ago—a second pair of stones was introduced by making the mill large enough to accommodate another driving wheel farther back on the wind-shaft, with another stone-spindle. This became known as the 'head-and-tail' post-mill. At this time, watermills, if they had two pairs of stones or more, still had a separate waterwheel for each pair of millstones. Smock-mills and tower-mills could not have a head-and-tail layout because the body did not travel round with the sails and wind-shaft, and at first they had only one pair of stones.

Intermediate gearing, where a main upright spindle is geared to more than one stone-spindle through spur wheels, came later; either the economic advantage of using one large waterwheel instead of two small ones, or (in windmills) the need for some means of driving two pairs of stones in a smock-mill or tower-mill, led to the innovation of laying two pairs of millstones side by side, each pair driven off a single large spurred wheel on the centrally placed upright shaft. To distinguish post-mills of this type from head-and-tail mills, they are called 'spur-gear' mills; but as all smock-mills and tower-mills are necessarily on this plan, they are not specially designated in this way. The wheel engaging with the brake-wheel, then, was not a 'stone-nut' (the name for a pinion on the millstone spindle) but an intermediate gear, and it became known as the 'wallower wheel'.

Between 1790 and 1840, in the Midlands, a number of post-mills were shifted by dismantling and re-erecting, and to make up for the inevitable weakening of the rebuilt structure, the millstones were then moved down from the upper floor to a massive bench or table called a 'hurst-frame' on the lower floor; these are often referred to as hurst-frame mills. The hurst-frame was also now and then—but rarely—employed in smock-mills, e.g. at Alderton smock-mill.

As the hurst-frame was always in the breast or forward part of a post-mill, the millstones were side by side, in spur gear.

MILLSTONES

The very earliest grinding stones were made of local stone where any was available. A local stone is still found in use in Wales and Anglesey. As windmills and watermills developed, a suitable stone became standard over a larger area; the Derbyshire Peak stone was used throughout England, and indeed in many other countries, and the French *buhr* or burr-stone for grinding flour was also known in Europe and elsewhere (Pl. 16c).

Peak stones are formed from a single block of millstone grit; but they do not grind wheat satisfactorily, and an old miller told me that he had found that bran could not be extracted in large flakes from Peak-stone grinding—he thought, therefore, that flour-dressing and bolting machines resulted from the introduction of the hard burr-stone—a kind of quartzite—which left the bran flakes in such condition that the flour could be conveniently separated from them in a flour dresser.

Burr-stones, being built up of some fourteen pieces cemented together (see Pl. 16c), are not perfectly balanced on completion, and

13

it was found that if they were balanced in the standing state by inserting fixed lead weights on the lighter side, they still might not maintain a true balance in the running state. Henry Clarke patented an improvement consisting of adjustable balancing weights in 1859.

Burr-stones are not unsatisfactory for general grist work and are consequently found in use for this purpose in many a windmill or watermill that has ceased to grind wheat for bread. A burr-stone and a Peak stone together are serviceable also. Composition stones of an emery or similar mixture have come into vogue during the past century, and are favoured by some windmillers, especially for driving at a good speed with an auxiliary engine, as they grind faster (particularly with 'sickle' furrows instead of straight ones) and do not require the fine grooving or 'stitching' which is essential to burr-stones and Peak stones. 'Cullin' stones, made from material found in the Cologne district, are also sometimes used.

TENTERING GEAR AND GOVERNORS

Before the invention of semi-automatic governors, tension of the stones was controlled by raising or lowering the runner-stone by hand lever all day long, a 'tenter-boy' often being employed to do this, to the miller's instructions. J. Ferguson (*Lectures on Select Subjects in Mechanics, etc.*, 1776), said this was also attained by slipping wedges under the end of the bridge-tree on which the stone-spindle stood, presumably in the absence of a tenter-screw. Later, a now obsolete device was introduced. This was called a lag governor or inertia governor. Two balls were suspended from the ends of a rod projecting a foot either side of the spindle. As the spindle gathered speed the balls lagged behind, since they were prevented from swinging out centrifugally (see Pl. 16d). Another rod was attached to the usual collar for the steelyard, which raised or lowered the bridge-tree and so varied the tension of the stones (see Fig. 5). This rod passed through two eyes formed in the pendulum rods just above the balls, and so was caused to rise as the balls lagged, thus achieving a general effect similar to the centrifugal governor. This arrangement presumably served its purpose in one Derbyshire wind-

14

mill from 1794 until it ceased work in the present century, and was also to be found in several mills between this point and the east coast within recent years.

Centrifugal ball governors, as we now know them, are thought by some to have originated in windmills, but J. Scott Russell (*Treatise on the Steam Engine*, 1846) stated that Christiaan Huyghens first applied them to clocks about 1657. Clocks and mills (it is worth noting) were the only examples of complicated mechanisms of the seventeenth century. Thomas Mead patented the centrifugal governor for controlling sails, millstones, or dresser feed, in 1787, and Stephen Hooper in 1789. In 1825 Dr Alderson, President of the Hull Mechanics' Institute, read an address to the members in which he said (rather unkindly): 'I do not give Mr Watt any credit for his governors, or centrifugal regulators of valves, as some have done. The principle was borrowed from the patents of my late friend Mead, who, long before Mr Watt had adapted the plan to his steam-engine, had regulated the mill-sails in this neighbourhood upon that precise principle, and which continue to be so regulated to this day' (*Mechanics' Magazine*, 17 September 1825). The balls of a centrifugal governor fly outwards as speed increases, and thus raise the collar and the steelyard.

FLOUR BOLTERS AND DRESSERS

A most important development was the dressing of flour meal after the grinding; and Nicholas Boller or Bolter of Saxony is said to have originated the process of 'bolting' or putting the meal through a sieve.[1] The French introduced separation of the flour adhering to the bran after bolting; and Edmond Brunt provisionally patented a dressing and bolting machine in 1614, but did not bring it into production (*Abridgements of Specifications Relating to Grinding Grain and Dressing Flour and Meal A.D. 1623–1866*, 1876). Flour seems to have been spelt 'flower' up to 200 years ago.

---

[1] In earlier days the miller used a sieve called a 'temse' (Anglo-Saxon *temes*) which he reciprocated on a wooden frame that became hot with the friction— hence the phrase 'Set the "Thames" afire'. Temse-bread was made from flour better sifted than the average. Some grist millers still possess a little wickerwork sieve or 'temse' on which they prepared sufficient flour for domestic use.

In 1765 John Milne patented a machine having several grades of cloth or wire mesh to separate first-, second- and third-grade flour from bran, or to separate 'sharps'[1] from bran. A machine for making bolting cloths was patented by Benjamin Blackmore in 1783. In 1850 James Bell removed all the remaining flour from the bran, pollards and sharps by driving the 'stock' against the walls of a vertical bolting sleeve (the 'centrifugal' bolter); and in the same year (1850), Joseph Foot patented a means of making the sleeve of strengthened silk instead of wool; calico was also much used prior to the introduction of silk.

Many devices were patented during the nineteenth century for ventilating and cooling the millstones, either by withdrawing air from the stone-case by suction or blowing air into the eye of the stone with a fan drive, but this practice was only occasionally resorted to in windmills.

Machines were also invented for automatically re-dressing the grinding face of the millstones, but there is a limit to what a simple and reliable machine can be made to do, and the average miller preferred to achieve the result he wanted by painstakingly plying the mill-bill inch by inch over the stone (Pl. 28 *b, c*).

LOCAL DESIGNATIONS

Windmill nomenclature varies in different districts: post-mills with fantails were at one time called smock-mills in Sussex, smock-mills proper being called 'eight-post' smocks; in Suffolk they always refer to 'post-mills', 'smock-towers' and 'tower-mills'; and in Holland any tower-mill with 'battered' or sloping walls is a smock; only cylindrical walled towers are called tower-mills. Squat cement-faced tower-mills were occasionally called smock-mills in Buckinghamshire. Sails are usually called 'sweeps' south of the Thames; the fantail is the 'fly' in East Anglia and the Midlands, and a post-mill body is the 'buck' in East Anglia, the brake is the 'gripe', the brake-lever the 'fangstaff' and so on. (A glossary of technical terms and local names will be found at the end of the book, pp. 155–68.)

[1] See Chapter 7, p. 109, 'Windmill Flour'.

Chapter II

DESIGN AND CONSTRUCTION OF
WINDMILLS

✤

IN the following pages I shall deal in special detail with post-mills, the only buildings designed to move bodily and completely in the course of their everyday work. Smock-mills and tower-mills will afterwards be treated a little more briefly. In this chapter and the next, the reader will be repeatedly referred to Figs. 2 and 3, in both of which the same sixteen main items of structure will be found numbered 1–16. Fig. 2 shows a typical East Anglian post-mill, and Fig. 3 is a complete diagram of Brill windmill.

THE SUBSTRUCTURE OF A POST-MILL

We shall refer now to the simpler diagram (Fig. 2).

Post-mills are mounted upon four (or sometimes six) piers of stone or brick (1); these may run to anything from 1 to 12 ft. high above ground, and are about 2 ft. wide and 4 or 5 ft. long; the piers must be of good-quality brick or stone, well cemented; a cracked or crumbling pier will endanger the whole structure. One pair of opposite piers is normally higher than the other pair, so that the timber cross-trees (3) can pass over one another at the centre.[1] The mill revolves about the main-post (2).

The main-post, usually a great baulk of oak, 2 ft. 6 in. square, 18 ft. high, and weighing about 1½ tons, stands above the cross-trees and is quartered over them but is *not* resting upon them, as commonly supposed. Nor are the cross-trees secured to the brick piers in any way: they rest upon the piers under the great weight of the windmill, which even prevents them from shifting during a storm.

---

[1] Very exceptionally cross-trees are curved over and under one another and then all four piers will be of equal height.

They ought to be of oak, about 20 ft. long, 12 or 14 in. deep, and 10 or 12 in. thick, and should have about an inch clearance between them where they cross at the middle; this clearance is for circulation of

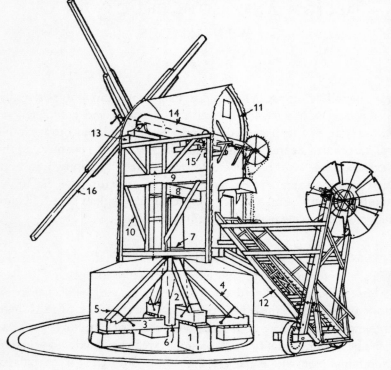

Fig. 2.   A typical East Anglian post-mill.

1. Brick pier	7. Centering wheels	12. Steps or ladder
2. Main-post	8. Crown-tree	13. Weather-beam
3. Cross-tree	9. Side-girt	14. Wind-shaft
4. Quarter-bar	10. Brace	15. Tail-beam
5. Retaining strap	11. Cap-rib	16. Sail-stock
6. Heel of main-post		

the air, and to allow for settling. (Brickwork weighs on an average 1 cwt. per cubic foot, and a good cross-tree weighs about half a ton.)

Holding up the main-post are the four oak quarter-bars (4), 10 or 11 in. square and weighing about 3 cwt. each. These are tenoned and housed into the main-post at their heads, and into the cross-trees at their foot, but not dowel-pinned or bolted, although often held at the foot by $2 \times \frac{1}{2}$ in. iron straps (5) passing over each quarter-bar and bolted with a 1 in. bolt through the cross-tree; this checks a tendency of the mortises in the cross-trees to split under the outward thrust exerted by the weight of the mill. The outer ends of the tenons of the quarter-bars should be set at right angles to the post or cross-tree into which they are socketed, and this face of the tenon should be 5 or 6 in. square, to give a sturdy bearing to take the thrust; and the flank of the tenon, tapering back in a curve, should be 9–12 in. long.

The head of the quarter-bar ought to be recessed an inch into the main-post, to prevent rain from draining into the joint; but the foot should overlap the cross-tree, and it must have two 'bird's-mouth' cuts across its width, either side of the tenon (technically called a double abutment and tenon: see Figs. 2 and 3, and Pl. 12*a*), to spread the outward thrust over the 'tree' as evenly as possible. When Bozeat Mill, Northamptonshire, was blown over, the cross-trees, which had seemed to be sound, were found to reveal cavities due to damp decay almost big enough to take a bag of cement; these were on the undersides, where the cross-trees rested on their respective brick piers. I believe the provision of 6 in. oak slabs on top of the piers helps to prevent such serious deterioration.

(When six piers are employed there are three cross-trees. Stokenchurch Mill may have been the earliest example of the kind in England. She was erected by William Winter in 1736; she was certainly the prototype for two or three new and reconstructed post-mills in Oxfordshire and Buckinghamshire, and only three isolated examples are recorded farther afield.)

At its foot the main-post is centred and steadied by its horns or tongues (6) which are an easy fit over the cross-trees; the base of the post is an inch or two above the cross-trees; the floor of the mill body is usually steadied against the post by means of inset wooden blocks known as 'woodwears' (Pl. 12*b*), or sometimes by an annular packing of tarred rope, but at Brill (Fig. 3) there are four large iron castors (7) running against an iron collar.

The top of the main-post is the chief bearing, about which the mill turns; it terminates in a pintle, onion-head, or cock-head, 8 or 9 in. in diameter and 7 or 8 in. deep, turned out of the solid wood of the post.[1]

Generally the main-post is iron-bound around the shoulder, and a thick annular iron plate (or washer) may encircle the pintle; or in old mills a wooden shoulder-piece 18 in. in diameter and 2 in. deep is made integral around the pin. This was perhaps preferable. At any rate on getting the crown-tree of Ashfield Mill, Framlingham, to the ground and turning it over, a few years ago, we found that the original iron ring on the shoulder of the post had become embedded in the annular recess in the crown-tree, and had then broken into a dozen pieces from constant pressure and friction. The socket had become enlarged and the pintle showed signs of rotting. A similar pintle at Bozeat Mill, Northamptonshire (blown down early in 1949 owing to the breaking of a quarter-bar) was still as hard and sound as a bell, even the bevelled edge being sharp and true, and the socket was scarcely worn, though no iron ring and no grease had been used. This post has the date beautifully carved upon it in period numerals —'R.H. 1761'; but was stolen from the mill paddock at Bradwell, Buckinghamshire, together with the crown-tree and the wind-shaft. In my opinion the crown-tree should be assembled without iron or grease, which seem to serve no useful purpose if the mill is properly designed; but Brill crown-tree is drilled for oiling the pintle.

In Sussex, where oak eventually became very scarce, four-piece pitch-pine main-posts were built up, rounded and slightly tapered in shape so that several $2 \times \frac{3}{4}$ in. iron hoops could be driven down from the top, and 1 in. bolts and $1\frac{1}{4}$ or $1\frac{1}{2}$ in. dowels were inserted to draw the sections tightly together. This was quite a sound method, offering the usual advantages of bringing face to face the reversed grain of the timber (as in halved and reversed floor-beams). Crown-

---

[1] The wooden pintle may also be inserted into the post, instead of being made integral with it, or an *iron* pintle may be inserted in the form of a cross-tailed gudgeon as described in the next chapter. The iron pintle at Rock Hill Mill, Sussex, was $5\frac{3}{4}$ in. high and 6 in. in diameter, and was cast integral with an 18 in. diameter cap to fit over the post.

trees down-bearing on the pivot of the whole mill were similarly constructed in the same county. At least one example survives, at Clayton, near Brighton.

THE BODY OF A POST-MILL

Of about the same cross-section as the foot of the main-post is the crown-tree (8), weighing a ton or more, resting upon the shoulder round the pintle. The pintle is centered in a socket cut in the underside of the crown-tree, about 1 in. deeper than the pintle itself so that the weight is not carried upon the top of the pintle. If it were, the turning of the mill would eventually wrench off the pintle and a crash would be imminent. When excessive wear occurs, so that the mill 'kicks' in a gale, an iron casting 1½ in. thick is fitted beneath the crown-tree, and its flange is made to make contact with a cast-iron collar or 'corset' clamped round the head of the main-post, the whole device, weighing perhaps a quarter of a ton, being called a 'Samson head' (Pl. 19a).

Crown-trees are of oak; or occasionally of pitch-pine or even ash; but if not of oak they should be quartered and reversed as with the built-up main-post described above. Each end of the crown-tree supports the side-girts or sheers (9), which are sometimes called the 'bressummers'. These side-girts measure 12–15 in. deep and 9 in. thick, and weigh about half a ton each. Side-girts should not be tenoned and mortised (though in Sussex and Oxfordshire they were) or they will eventually crack at the middle, and the mill will go to pieces. They can rest on an inwardly inclined rebate on the crown-tree, and can be secured there with dog-irons or rods and bolts. To the ends of the side-girts four vertical corner-posts (9 in. square) are mortised and dowel-pinned. The rest of the framework and floors of the carcass of the mill is to be built round these timbers, usually with suitable diagonal braces (10) to maintain the rectangularity of the body. But Moreton Mill, Essex, and Six Mile Bottom Mill, and Bourn Mill, Cambridgeshire, have no side-girts; the load is communicated to the crown-tree through upright posts supporting extra heavy top rails. They are dated 1715, 1764 and about 1636.

Post-mill bodies have a floor-space of 16 or 18 ft. by 10 ft., although a few are as much as 20 ft. by 12 ft., and they measure from 22 to 25 ft. from bottom floor to roof ridge; heavy iron tie-rods or stay-rods with turn-buckles are frequently provided (as shown in Fig. 2), particularly in Suffolk, where post-mill construction was better understood than elsewhere. One pair of these stay-rods can strike down vertically through the side-girt and crown-tree to the bottom frame of the body, and one each side of the main-post from crown-tree down to the two sheer-trees that flank the waist of the main-post underneath the lower floor. The stay-rods are intended to help to carry the weight of the body and the machinery.

Unless they are thus supported, the sheer-trees, which are about a foot square, and run the full length of the mill, should rest above a collar or girdle, a heavy square wooden frame gripped around the main-post immediately over the heads of the quarter-bars. Normally the 'collar' steadies the mill body in rough weather, when undue strains would otherwise be thrown upon main-post and stairs by the thrust of the wind on the head of the windmill. Should the side-girts crack or weaken, the collar affords support to the sheer-trees and framework, thus relieving the side-girts.

In the Midlands, although the body may measure up to 18 ft. long by 12 ft. wide, no diagonal braces or rods are used; the mortising of the framework and the intermediate studs is relied on to keep the body square; not unnaturally, on some of them the head eventually sinks badly. They often have an 'inter-tie' in the lower side wall, between side-girt and floor.

Cap-ribs or rafters (11) spring from the upper side-rails or head-rails of the mill to meet at the roof-ridge, where they are overlapped; and over the whole exterior is the weather-boarding, of chosen quality (7 × ¾ in.), well lapped to exclude driving snow, and advisably running the full length of the mill to avoid joints. A 'storm-hatch' or 'weather shutter' is always provided above the wind-shaft for tending the sails, etc.

That part of the mill ahead of the main-post is called the head or breast; and rearwards is the tail, with a large door for the miller to be able to get millstones in and out. Sacks are sometimes handed out of the doors and slipped down a sack-slide on the ladder (12), but Brill Mill, here illustrated, had no slide in recent years.

Except in East Anglia, the steps are coupled to the bottom rail or sill of the body (which also constitutes the threshold of the doorway) with a long bolt and 'eyes', like the tail-board of a cart (Fig. 3), but a better way is to anchor the side strings of the steps into special double doorposts, which allows a very useful platform to be extended from doorway to steps, with a canopy over it (Fig. 2). The side strings of the steps may be up to 12 × 6 in. section. On the tallest post-mills the steps weigh half a ton, and have 40 or more stairs, with diagonal tie-rods and timber braces underneath. Ties are particularly necessary if a fantail luffs the mill by driving the steps round.

Across the head of the mill is the weather-beam (13), sometimes known as the breast-beam, about a foot square. The weather-beam supports the head of the wind-shaft (14), and is secured to the top side-rails or head-rails of the body (at the eaves) by dog-irons and short tie-rods, with additional support from the 'prick-post' in the breast (Pl. 19b), to which it is fastened by iron plates.[1] The recoil thrust from the driving gears, coupled with the weight of the sails, tends to wrench the weather-beam forwards and downwards; there is also a tendency for the roof to creep forwards, and this may be checked with long 1 in. iron straps (coach hooping) attached diagonally across the ribs. Likewise, the two upright weather-studs one each side of the wind-shaft (they should have big wooden cheeks bolted to their inner faces) have to be held back by weather-irons (or 'keepers'), where they stand on the weather-beam; and as the gable of the cap tends to be wrenched over in the direction in which the sails revolve, the weather-stud on that side is braced with one or two struts (Pl. 20a).

[1] An important distinction should be made between three types of breast walls—in East Anglia there is a 'prick-post' and four diagonals; in the Midlands three or four transoms with numerous upright studs between; and in Sussex a 'prick-post' with numerous transoms either side of it. A lot depends on the breast; it supports the sails.

ROUND-HOUSES

A round-house is not an essential part of a mill but an addition for protecting the timbers of the substructure, the cross-trees and quarter-bars, and providing storage space. Round-houses are usually built of 9 in. brick or stone, with tiled or wooden-boarded roofs. But in Sussex wooden-walled round-houses were popular, and the wall often terminated 2 ft. above the ground, the floor beams resting on little brick piers; this ensured thorough ventilation.

Room enough is left round the main-post for sacks to be lowered into the round-house from above; and if there is an intermediate floor it is provided with four sack-traps to allow for the mill turning in the wind. It should be noted that the circular wall of the round-house in no way serves as a support to the mill, except, rarely, where the round-house has been provided with a curb to receive a set of travelling wheels mounted on a substructure under the body.

But a rare practice, favoured on the Oxfordshire and Buckinghamshire border, was that of mounting a strong timber curb upon the quarter-bars to carry some of the weight of the mill on castors. This curb came within the round-house roof (as still existing at Chinnor), but was no part of the round-house structure. A similar device was incorporated on the neighbouring Bledlow Ridge Mill (Pl. 2). Both mills had three cross-trees and six quarter-bars. Possibly the third cross-tree was there to give better support to the curb and to distribute the strain.

Two doors are provided to a round-house, because the sails, which might be over one door or the other, will not usually clear a cart, and may often reach within a foot or two of the ground.

SMOCK-MILLS

Smock-mills (originally called frock-mills) must have got their names because their white-painted walls and tops suggest a smock-frock and cap. They are usually octagonal (but occasionally hexagonal or ten- or twelve-sided) with oak or pitch-pine cant-posts (about 9 in. square in section), at the angles, connected together by a bracing of horizontal struts (transoms) and crossed or herring-bone diagonals covered with weather-boarding. The cant-posts of a smock-mill (which are always canted inwards) stand on a timber sill secured to wall plates in the foundation brickwork.

A weakness of all smock-mills comes from the difficulty of fixing the cant-posts satisfactorily to the sill. Often they are strapped with iron and bolts to oak blocks on the inside of the sill (this also serves to secure the joints in the sill itself), to counter the tendency for the feet of the cant-posts to splay outwards, and eventually to slip off the brickwork, which would bring about the collapse of the mill.

Another method was to mount the posts in big cast-iron shoes bolted down upon the sill; but neither arrangement is proof against rotting of the feet of the cant-posts, which has prevented most smock-mills from surviving as long as 200 years. Lacey Green Mill, which dates from 1650 and is the earliest smock-mill known, was brought from Chesham in 1821; and although the posts appear to be the originals, I suspect that they were renewed when it was re-erected; the mill would then have had something like a fresh start, and has thus been able to outlast hundreds of newer smock-mills.

The windmill enthusiast is apt to pay slight attention to the framing of a smock-mill body, perhaps assuming that it is bound to follow the same principles of construction as a two-storey barn, though he would think it his duty to make a close inspection of the framing of a post-mill. But the smock-mill is in fact a very interesting and instructive piece of work, as can be seen at once from the photograph of Mr Denis Sanders's model of Shiremark Mill, Surrey, in Pl. 13 *b*. Usually the two main beams of each floor are partly housed in the cant-posts, and partly resting on the two transoms at opposite sides of the mill. The other six transoms of this floor are set slightly higher —level with the upper face of the beams—so that they may carry the ends of the floor joists; the transoms are usually strengthened by intermediate upright posts linked with the cant-posts by diagonal braces. Lacey Green Mill, however, does not conform to such set rules of construction. Main beams ought to run opposite ways on alternate floors; and an enclosed hurst-frame or cog-box carrying the driving gears is sometimes mounted between four upright posts each about 9 in. square, placed at the intersection of the beams of the respective floors.

Brickwork may form only a foundation at the ground, or it may be carried as many as three floors up, forming what is known as a 'roundel'. Round the top of the mill, if it is finished in timber work, is a circular curbing of wood, fixed with iron plates, with sometimes an iron curb superimposed. (In brick or stone-built tower-mills this curb is secured by four or more very long and substantial stay bolts reaching down 10 or 15 ft. into the brickwork, which may have to be chopped away to get at the bolts if the curb ever needs renewing.)

Both smock-mills and brick-built tower-mills can have up to six floors (even ten or twelve in a tower-mill). A common arrangement is to have a 'dust' floor below the cap, where dirt and wet always penetrate. Below this is the bin floor for grain, then a 'dresser' floor for the flour machine, next the stone floor, followed by a meal or sack floor. This may be the ground floor, or the ground floor will be reserved for storage and loading.

The sack-hoist is commonly on the dust floor, which was called the lantern floor in the days of the lantern-pinion 'wallowers'.

A few smock-mills have been rebuilt as tower-mills; they may be recognized by their octagonal brickwork at the base, with circular upper walls.

TOWER-MILLS

Tower-mill construction, although more orthodox from a builder's point of view, must have had its special problems. In any event, post-mills have outlasted smock-mills and tower-mills of similar age. Excepting Lacey Green's reconstructed smock-mill, all the dozen or so seventeenth-century windmills surviving in a more or less complete state in England are post-mills; and in Suffolk, where post-mills and tower-mills were equally well-built, six post-mills and only one tower-mill were working by wind in 1948.[1] In East Sussex, where

---

[1] The local millwrights regard post-mills to be three times as safe as smock-mills or tower-mills, because of the greater chance, they say, that the sails may be dislodged from smock-mills and tower-mills in a gale.

smock-mills were popular, the millers say they never cost less than £1000. A high cost is itself no recommendation for what is a less efficient structure than the post-mill; post-mills were the cheapest; tower-mills were the dearest.

Nevertheless, a number of modern tower-mills, about a century old, more or less, are still in good working order, notably in Lincolnshire; and Cranbrook Mill, Kent (Pl. 8), which is probably the best and most expensive smock-mill ever built in England, is in excellent trim.

Brickwork of tower-mills should be at least 18 in. thick—preferably more—up to the stone floor, so that there is no need for thickening or corbelling out under this floor, as is found in the Kentish-built mill at Coleshill, Buckinghamshire; the upper walls are often made half the thickness of the lower walls, sometimes with an intermediate thickness half-way up. Good foundations are essential; brickwork weighs nearly a hundredweight per cubic foot, and the total weight of a big mill (almost 500 tons for Wendover Mill, which has exceptionally thick walls) imposes a great strain on its restricted ground space. A load of 2 tons per square foot of foundations should not be exceeded on average building land; this is about the loading at Wendover, where the walls are 3 ft. thick and 82 ft. in outer circumference.

Several Norfolk and Suffolk tower-mills were built 100 ft. high or thereabouts; and Southtown Mill, Gorleston, 122 ft. high and 42 ft. in diameter at base, was put up in 1812 at a cost of £10,000. It carried 84 ft. sails. It was surmounted by a lantern and telescope.

Mr Paul Baker, the Nottingham windmill expert, found two six-sailed tower-mills in one mill-yard, although without their sails and machinery, at Leabrooks, Derbyshire; these were 65–70 ft. to the curb and more than 35 ft. in diameter at their bases; they were built in 1877. They retain their ogee caps (Mr Baker is endeavouring to arrange for the preservation of these before they go to pieces), and both have an iron staging bolted to the outer walls, just below the curb .The six-armed iron crosses for the sail-arms, now lying broken in the yard, were far enough out from the wall to clear a narrow stage, and so to permit of walking behind them.

Octagonal tower-mills need very good bonding of the brickwork at the corners (as do the brick bases of smock-mills), and they should have the individual courses of the bricks inclined inwards, so that the upper face of the brickwork is perpendicular to the cant or 'batter' of the wall—if the bricks are laid in a horizontal plane, as in some East Anglian smock-mills, they are then necessarily slightly stepped all the way up, and need to be coated with tar to keep out the wet. But all tower-mills, unless of exceptionally good quality brick, benefit from a dressing of tar, or a rendering of cement, though they lose something in good looks by this treatment. Windows ought not to be vertically underneath one another (though they usually are), as this increases the tendency for the walls to crack. Floor beams should run a different way for each floor; and should rest on oak blocks or templates in recessed pockets in the walls, as at Quainton Mill, Buckinghamshire; sooner or later wood in contact with brickwork will rot. For this reason, too, it is not good practice to fix timber wall-plates in the walls for attaching a wooden-boarded lining to the stone floor, or for other purposes.

THE CAP OF A SMOCK-MILL

This may be of the hooded or wagon-top pattern, or of similar construction to a post-mill roof, or boat-shaped, like a short inverted rowing boat; or it may be a simple ridge-roof, an ogee-shaped cupola finished with a 'dolly' or acorn-shaped finial, or even a conical or elliptical cap; but it should preferably make some show of streamlining so as not to impede the air flow behind the sails.

The cap-ribs may be of poplar, steamed to shape, or of some wood that has grown with a suitable curvature. Close-fitting rounded caps generally have a skirt of upright boarding surrounding the curb to keep out the weather; they can be vertically boarded, like a barrel, or finished with horizontal weather-boarding.

Messrs Saunderson's elegant ogee caps in Lincolnshire (Pl. 27*b*) have both horizontal (inner) and vertical (outer) boarding outside the cap-ribs; and the upper ends of the ribs are housed into the face of a wooden boss, upon which the upper ends of the vertical boards are also nailed. Above the boss is the wooden stem of the 'stalk-and-ball' finial, with a two-piece cast-iron globe at the top of the stem; and a vertical tie-rod passes centrally through the whole assembly. (More stumpy finials elsewhere have the ball and stem fashioned from a single baulk of wood, with a lead cap on the ball, and no iron-work.)

It is desirable to give what the building trade calls a 'marouflage' finish to ogee-pattern caps. To do this a covering of sail-cloth is applied by coating both it and the boarding of the cap with a mix-ture of gold size, varnish and white lead paint. This is made up to the consistency of a thick cream, and applied twice if necessary, at an interval of three hours, and the cloth is then spread and tamped down over the boarding while the adhesive is tacky.

Some tower-mill caps were copper-covered—a very good prac-tice; but when galvanized iron came in, over a century ago, it was adopted not only for new caps but also even substituted for the copper, which was ripped off and sold for its value. Alas! sheet iron always pulls the nails out of timber owing to temperature changes, but copper fits down like a glove indefinitely; it is safe to say that iron ought never to have been used unless the sheets were bolted, riveted, or welded together and finished with bituminous paint, which they never were. Instances occur (chiefly in Lincolnshire and Sussex) where the actual cap frame, and even the fan-stage, is entirely of iron; but wood has always been the material primarily associated with windmills, and the millwrights will ever be famed for their skill in using wood for the job.

CAP DETAILS AND MECHANISM

With tower-mills and smock-mills the cap is that part of the mill which is movable to meet the wind. Around the base of the cap is the cap-frame, running upon the curb-ring of the mill; and the two, if of wood, may run face to face, well greased, and usually with iron plates bedded in the underside of the cap-frame; when the cap-ring slides upon the curb the mill is said to have a 'dead' curb. 'Live' curbs travel on weight-bearing iron trolley-wheels mounted underneath, or let into, a cap-circle, and there are always centering or truck-wheels to maintain the cap centrally upon the curb. If there is an iron curb, the rims of these truck-wheels often fit beneath an over-hanging flange to prevent the rising of the cap in a gale, but they are for the most part kept free from contact with the curb to prevent unnecessary friction and binding. Indeed, the actual weight itself of a massively built cap of large diameter appears to serve as check enough.

Generally the cap-frame is constructed about a pair of longitudinal main sheer-trees from 6 to 12 ft. apart, with three or four cross-ties, one of which, at the centre, is the sprattle-beam (which carries the bearing for the head of the central upright driving shaft); another, farther back, is the tail-beam for supporting the end of the wind-shaft. These timbers are very sturdy, about a foot square in section; and into them are mortised and housed some outriggers, from which, in East Anglian mills, the centering wheels project on iron brackets, or in Sussex and elsewhere, on large wooden blocks.

Some northern tower-mills, of which the big Upton Park Mill, Chester, is an example, are furnished with a centering frame under-slung from the cap-frame proper and designed to encircle the main upright shaft beneath the wallower wheel; this would effectually fore-stall any lifting of the cap. Elsewhere a cap-circle in the form of an iron ring may be affixed beneath a heavy square frame which serves as the main foundation of the cap structure, dispensing then with the sheer-trees. Wendover and Upminster mills have a floating roller-bearing called a 'shot-curb', between mill and cap.

More trolley-wheels are provided at the head than at the tail, because of the weight of the sails; and more truck-wheels at the tail, to withstand the thrust of the wind upon the sails and the cap gable. The thrust of the driving gears (which tends to thrust the weather-beam forward while the sails are actually driving the mill) may be checked by strong iron straps carried back from the weather-studs to the main cap sheers or to one of the cross-ties.

GOOD ENGLISH OAK

All materials used for windmills by established millwrights were chosen of the best quality. Careful selection of the tree from which (say) such an important item as the main-post was to be fashioned was essential; for not all oak is good and immune from breakage. 'Do not take a tree from a wood on poor soil', said the experienced millwright; 'and beware of black Italian oak, which is heavy but not always suitable—choose a good English tree growing in a hedge-row on strong heavy stony soil.'

Years ago when oak trees in some districts were barked some months before felling (the bark being highly valued for tanning), it was generally acknowledged that a tree barked in springtime and felled in October provided the toughest all-round material and was least susceptible to attack from the worm. By this process the sap was released in the spring, the remainder of it being taken up by the growing foliage during the summer, so that by October the sap-wood was nearly as hard and firm as the heart. Moreover, autumn-felled timber dried slowly and naturally, with least risk of splitting; and it was even claimed that by squaring up the timber instead of leaving it in the round, and drilling it from end to end (as for a 'patent' sail wind-shaft) it would dry out still more evenly; but of course such details will not be attended to if a timber dealer's interest is only in quick sale and maximum profit.

Oak should be felled at about 100 years old. It is not hard and mature enough at 60 or 70 years (as now cut), and at 150 years it may be getting comparatively brittle and lifeless, being by then beyond its prime. By observance of these standards (born of long experience in a more leisurely and efficient age), windmills were devised and erected which would stand for centuries, monuments of carefully applied knowledge and expert craftsmanship (see Tredgold, *Carpentry*, 4th ed., 1883, pp. 336–41).

Wheelwrights, on the other hand, who built huge and beautiful farm wagons to last 60, 70 or 80 years, often preferred spring-felled oak in districts where trees were not barked before cutting. Autumn-felled timber was comparatively rare and expensive in such districts, because of the great demand for spring bark, and oak (for wheel spokes) is more readily cleaved with the axe if cut in springtime. Wheelwrights also liked the timber to develop limited cracks or crevices across the grain. They were convinced that without these 'shakes' (which the millwright did not welcome) timber would not stand the incessant bumping and jolting to which farm wagons were subjected on unmade roads and tracks (G. Sturt, *The Wheelwright's Shop*, 1948, pp. 24–5). Thus in days gone by every man looked with the eyes of his own trade. Today a piece of oak is—well, just 'a piece of oak'. A man today may even look at a bad piece of oak in a house, and say 'Well, of course, it's only English oak'; and then I am tempted to ask 'Since when has English oak been inferior?' Certainly not when windmills were being built.

All the main timbers of the mill body can be of oak; or they may be pitch-pine, or even sweet chestnut, although curved cap-ribs can be elm, poplar or ash; oak dowel-pins were used, in conjunction with hand-made spike-nails, strap-bolts, dog-irons, coach-screws and bolts, and iron plates. Floor boards had to be of best 1 in. well-seasoned timber, double-ploughed and tongued, so that crevices where meal would lie and go sour or trickle through to the floor below might be avoided.

PROTECTIVE COATING

Tar is the best preservative for post-mills and smock-mills, though, compared with white paint, it greatly detracts from the fine appearance, and is messy to apply. Creosote has useful preservative properties, but is less impervious to wet unless applied very frequently. Creosote can be sprayed on to the mill body with an ordinary stirrup pump, and looks better than tar on the sails. White paint is very expensive, even when the miller mixes and applies his own paint. Two or three hundredweight of paint is needed to do an all-white mill thoroughly, and if it is left until it begins to peel in the sun, the peelings will actually catch and lodge the wet, doing more harm than good. A tarred or creosoted mill can be painted white if first coated with red oxide; or new wood can be treated with Cuprinol (colourless) and either left thus, when the wood will presently turn a pleasant silvery grey, or painted white after an interval.

To keep the interior dry and free from draughts, and to facilitate sweeping down the walls, a few post-mills were lath-and-plastered within. Smock-mills occasionally had plaster-lined caps; and many, but not all, tower-mills are plastered internally.

Chapter III

THE MILL COMES TO LIFE[1]

❧

THE millwrights of the best days of mill-building undertook to build and make and install all the complicated strong machinery (see Fig. 3) in a cunningly balanced, freely revolving post-mill. This was before the days of general education, and sometimes without help from accurately drawn plans and diagrams; and their mills withstood the weather, did their work, perhaps for hundreds of years, and went out of use only for economic reasons. And now almost all of that exact skill and knowledge, with the passing of millwrights, is in danger of being forgotten. I hope my book, the result of a quarter of a century's much-enjoyed study of their masterpieces and methods of work, may do something to keep alive the affection and admiration of later men.

To understand the skill and ingenuity used by the old millwrights in contriving and building their windmills it is necessary for us to examine in more careful detail surviving examples of their structures, and here will follow some closer details of their work and methods.

But mills were built long before the notion of standardized or mass-production methods was born. Each surviving windmill contains in itself some unique solutions of the many practical problems that cropped up. To be an old millwright (one can say) is easy enough! All you want is a building (which you specially make) capable of meeting and dealing with wind from any quarter, and of converting wind-force into grinding-force. You must, of course, provide space enough within to house the grain, and to feed the grain to the grinding machinery, and to collect the meal, and to make saleable flour, which is to be got away from the mill to the baker, who will make bread of it—good bread, that goes without saying.

---

[1] Dealing particularly with Fig. 3.

So you must make a building that will meet the weather and withstand it, that will seek the wind and receive it and take rotatory power from it; that will house whirling machinery under the miller's exact control. You must find out the most efficient wind-motor, and make every part of it yourself—out of trees for the most part; and once your mill is made to stand, and to work, it must be safe, long enduring, flexible. The mill must take the wind so that the wind turns it; you must convey the turning motion to the machinery and, by cogs and axles and spindles and pivots, to the grinding-stones; you must collect what is ground, and by that motion dress it into the flour that men need; you must take in grain and send away flour. Have patience then, in contriving how to take a giant's power and use it; for you are not a giant, except in ingenuity and patience.

This is the story of the achievements of great men of past days, each learning from earlier men's experience; of how they thought to capture the wind, and harness it to human purposes; and how, in fact, they succeeded, as evidenced by the machines that have survived them. These survivals are no longer of great practical use to us; men have found other ways of grinding corn, faster and more profitably, but old men's skill and triumph with the wind-giant can still be comprehended and can inspire us with fraternal pride. Only, battered still by weather and time, the mills they built are slowly disappearing and will at last go out of mind, with all the skills they exemplify, if we do not now learn from them—and soon—what they and only they can give us of the greatness and skill of men of the past.

THE MOVING PARTS IN SEQUENCE

The wind on the sails revolves the wind-shaft. The brake-wheel on the wind-shaft is cogged and engages with the wallower wheel, which is mounted on a vertical pivoted shaft revolved by the wallower. A cog-wheel lower down on the same shaft engages with two stone-nuts (one on each side of the main spindle) and revolves these nuts, which are usually beneath the stone floor, as in Fig. 3.

From each stone-nut a vertical stone-spindle passes up into the central openings or eyes of its pair of millstones, and this spindle is so coupled to an iron bar cemented into the centre of the runner-stone that it will revolve the runner-stone above the fixed bed-stone. Grain, fed in at the eye of the runner-stone, is by the grinding action driven forward between the runner-stone and the bed-stone and is there ground to meal between the two.

There is space enough on each side of the iron bridge piece or bar to allow the grain to be fed into the eye of the stone. Much care is taken to ensure a correct balance of the stone when it is running, and adjustable lead weights are added to the stone to achieve this balancing.

Thus when all is in order the balanced runner-stone revolves close above the bed-stone, with grain passing between it and the fixed bedstone, being meanwhile reduced to meal. The runner-stone is revolved by its spindle, which is in turn revolved by gearing to the wallower shaft, and that in turn by the brake-wheel on the wind-shaft. Now we will consider these parts in detail, one by one.

WIND-SHAFTS AND BEARINGS

Mention has already been made (p. 23) of the wind-shaft and its supporting beam—the weather-beam as it is called—in the head of the mill. In ancient times, the heavy wind-shaft neck rested in a cup-like half-bearing (of wood, basaltic rock, marble, or stone) fixed on this weather-beam; and the wind-shaft was entirely of wood, with a number of iron fillets set longitudinally into the wood of the shaft at the neck, and separated by strips of wood, the whole secured with iron bands to form a bearing (Pl. 14b); later it became usual for wind-shafts to be provided with a solid iron head (if the shaft is not entirely of iron), and this head generally runs in a brass or gun-metal half-bearing set in an iron seating or pot.[1]

Many Norfolk windmills, and some elsewhere, also had the tail-end of an all-iron wind-shaft separately cast and joined on to the

---

[1] Occasionally this pot swings on trunnions resting on an iron pillow-block or bolster (Pl. 31 b) when it is called a 'swing-pot'; and the pillow is bolted to a heavy neck board mounted on the weather-beam. This keeps the brass in true alignment with the shaft, which is often not the case with a rigid bearing.

shaft, of which it was to form a part, with flanges and bolts. Why the wind-shaft needed thus to be constructed part by part we do not know now; if the poll-end and neck were similarly bolted to the remainder of the shaft (as at Bradwell, Buckinghamshire), it may merely show that this particular millwright lacked the workshop space for making single long castings.

Wooden wind-shafts taper, it may be from 2 ft. diameter at the head to 18 in. diameter at the tail. A long shaft probably weighs half a ton, plus another half-ton for the iron head known as the 'poll-end' or 'canister', if such is fitted.

Iron wind-shafts vary greatly in size. Some taper from 12 in. diameter at the head down to 5 or 6 in. at the tail; some are twice as long as others—say, 8 ft. to 16 ft. They may weigh anything between 1 and 5 tons, inclusive of head. Iron necks are 10 to 12 in. in diameter and 9 or 10 in. long, resting in a brass half-cup 6 in. deep and 2 in. thick. If an iron head is mortised into a wooden shaft the iron part will be some 6 ft. long overall, including the canister (or iron socket for the cross-arms of the sails), with four big long fins (like the tail-end of a bomb) inserted into the woodwork, with cross-bolts through them and several iron rings forced on cold, or straps clamped round the wind-shaft, to stop the wood from splitting (Pl. 14c, e).

Behind the canister is a groove to prevent rain water from running down the shaft; and at the tail of the wind-shaft there may sometimes be a heavy iron thrust-pad or ring. A cross-tailed gudgeon carrying a pin about 3 in. long is mortised into the tail of a wooden wind-shaft; but an iron wind-shaft would have an integral tail-pin.

The neck-brass or cup is made longest at its leading edge to stand up to the thrust of the revolving shaft, and is furnished also with a shoulder to rest on the edge of the bolster or swing-pot, and outside lugs to register with recesses therein, so that it cannot be pulled round with the shaft and so dislodged (Pl. 14a and 31b).

The tail-bearing ((15) in Fig. 3) can be a plain iron ring; in which case there should be an annular iron thrust-pad surrounding the gudgeon pin. But preferably the bearing should consist of a properly designed cup of wood or brass, seating the gudgeon, with provision for a flange at the tail-end of the gudgeon, and having a tail wall to the brass, of good thickness, to serve as a thrust-pad, the whole being housed in an iron pot. The iron pot in its turn will be mounted on a cross-member (the tail-beam) which is wedged into suitable guides (called upper side-girts) built into the walls (Figs. 2 and 3), so that it may be slightly moved a trifle forwards or backwards to allow for taking up the gradual wear of the thrust bearing. Any such adjustment *must* be made when the sails are in motion, or the shaft may wrench the weather-beam foward instead of riding forward in its brass seating.

Over the tail of the wind-shaft there is sometimes fixed an iron hoop or keep to prevent it from lifting in a gale. Such a thing may happen if the wind is upon the back of the sails; a mill is then said to be 'tail-winded', and is in danger of losing its sails altogether. A wooden keep is provided at Brill. A few wind-shafts have a massive iron counterbalance clamped to the tail, like a solid drum, to hold them down; and one or two have had a roller neck-bearing.

SAIL-STOCKS

Old-time windmills had the sail-stocks ((16) in Fig. 3) mortised through the head of the wooden wind-shaft (Pl. 14*d*), but this arrangement is rare, and the usual cast-iron canister-head, otherwise known as the poll-end or cross-eye (17), is more generally secured to a wooden shaft. The stocks are passed through the eyes (or boxes) of the canister and wedged, not bolted. Iron wind-shafts have the poll-end or canister cast as an integral part of the whole shaft, or a four-, five- or six-armed cross (according to the number of sails) may be keyed on to the head of the shaft (Pl. 24*c*). Mortised wooden wind-shafts (Pl. 14*d*) without a canister were serviceable for moderate-sized sails, and on marsh-mills, and did not tend to cut the sail-stock, as an iron canister may. To centre the stock in the poll-end or canister a shoulder is provided on the side of the stock. The stock should be of pitch-pine, anything from 30 to 60 ft. long according to the sail span, and 9–12 in. thick by 12 or 14 in. deep at the centre. Each shaft is tapered down to some 5 in. square at the tips. Oak sail-stocks were sometimes used in the Midlands, but they were

37

generally made too short (because of their weight), and it was diffi-cult to obtain a suitable length of oak without large knots. A very old miller in Sussex, however, told me that stocks had been of oak when oak was plentiful, and later pitch-pine stocks, or so he believed, were a substitute.

One of the disadvantages of oak is that even if it is used in the dry, the tannic acid that oak contains will quickly rust any ironwork in contact with it. Larch fir, which has a long grain, is fairly service-able, if a large enough tree can be found; Scots pine has a short grain, and is too liable to snap; Oregon pine is little better.

Hornbeam or apple-wood poll-wedges, with only a comparatively slight taper, should be used to wedge the stocks firmly in the head, with a 'stop' behind the wedges to prevent them dropping out if the wood should shrink. The chief drawback with poll-ends or canisters is that they collect the wet, which does not afterwards readily dry out; an idle mill therefore should be left with the shoulders of the stocks against the upper side of the canister, so that they do not lodge the water; this position would also prevent the stock slipping down should the wedges loosen.

A good practice, although discounted by some millers, is to strap a pair of long clamps of pitch-pine (Fig. 2), one on each side of the sail-stock, passing outside the poll-end, as a safeguard against the stock becoming cracked or rotten at the centre. Average clamps 20 ft. long would taper from about 7 × 5 in. at centre to 4 × 3 in. at the tips, and their inner faces should be straight so that the clamps will be strained up to the stock, leaving a ventilation channel from the canister along to the tip of the clamp; clamps should be shackled rather than bolted, as bolt-holes may enlarge by the oscillation of the clamps and every bolt-hole lodges some wet and so becomes in time a source of weakness.

SAILS

Most windmills have four sails (or 'sweeps', a Sussex term), but multi-sailed mills (popularized by John Smeaton) have been built, especially in Lincolnshire, where five-, six- and eight-sailers can be seen (Pl. 35). In emergency, four-sailers can run with two sails; sixsailers with two, three or four; and eight-sailers with two, four or six; five-sailers cannot be balanced without their full complement and are therefore at a disadvantage, though a few survive. So far as is known, all post-mills, with one exception, had four sails; so did smock-mills, with very few exceptions.

Arms, 'whips', or sail-shafts ((18) in Fig. 3), preferably of pitch-pine, are fastened to the stock with two or three through-bolts and two wrought-iron straps. The 'whips' taper from about 8 in. wide to 4½ in. on the face; and they are thickest, of course, where they leave the tip of the stock.

Sail-bars (19), 3 × 1 or 1¼ in. deal, are mortised through the whips; and for cloth sails, 'uplongs' (20) are fixed to the bars, with a 'hemlath' (21) at the outer edge, with back-stays behind the sail. One or two 3 in. wire nails are driven through where the uplongs cross the sail-bars, and then clenched; but the sail-bars are usually driven home in slightly tapered mortises, and not nailed. The tail-ends of the sail-bars, on the other side of the whip, support a leading-board or wind-board (22), the outer edge of which should be wedged up from the sail-bars, so that it directs the oncoming wind into the sail, and helps to hold the cloth firmly to the frame. This wind-board is always on the leading edge of a 'common' sail (as the cloth-covered sails are generally called), and the lattice frame and cloth are on the 'driving' side.

An iron rung (23) is provided at the inner end of the sail, near the poll, and on the sail cloth are several rings which will slide along the rung. Cloth, of the texture of rick-cloth, is secured to the whip at intervals and can be spread or reefed and furled, according to the wind, being 'sailed out' and secured in position by two or three long sail ropes or 'pointing ropes', which are passed round the sail frame and tied down to the tip of the sail-shaft (Pl. 23). The longest rope, attached near the heel of the sail-shaft, draws it out to 'sword-point' position, the next one to 'dagger-point', the third rope to 'first reef', and the cord at the tip to 'full sail'; as a rule the sails would all be spread or reefed alike; but in rough weather one pair might be completely furled and the other pair, say, at sword-point. When furled, the cloth is thrown round the cleat (24) and usually twisted

39

in and out of one or two of the sail-bars near the tip to prevent the wind from tearing it away. Such sails weigh something under half a ton each and, with 30 ft. stocks, the total weight of the sails is about 3 tons.

Nowadays, as the best materials for sail-stocks and whips are no longer obtainable, the writer favours a system of rigging with stranded aeroplane wire or ordinary fencing wire strained across between the tips of the stocks (or the sail-tips), to help the sails to support one another, with similar strainers from a projecting standard or bowsprit in front of the canister, to accept the wind pressure on the face of the sails. (An isolated example in England is found at Stanningfield (Bury St Edmunds). The rigging at Brill can be seen in Pls. 21 *b* and 23.)

The 'weather' or 'angle of weather' on the sail—that is to say, the twist in the sail, which becomes less pronounced towards the tip (diminishing from an angle of 20–25 degrees at the heel to 3 or 4 degrees at the tip, varying according to type of sail and ratio of driving gears)—indicates at a glance which way the sails run; also, the broad side of the sail is the trailing side.

More modern types of sails and luffing gear, and certain other details, will be described in Chapter 4.

THE TAIL-POLE

Old post-mills were turned or 'luffed' into the wind by a pole variously called the tail-pole, tail-beam, turning-beam, or tiller-beam ((67) in Fig. 3). The tail-pole had at its lower end two wooden bars forming a shoulder yoke, where the miller placed his shoulders, after first levering the steps clear of the ground by the lever or 'talthur' which he hitched under an iron peg. (The 'talthur' is seen on Drinkstone Mill (Pl. 13 *a*).) Sometimes he attached a horse or donkey to drag the pole round, or there might be a winch on the pole, from which a chain could be attached to successive chain-posts round the mill yard. Turning the winch would then haul the mill round to its next position. The pole was 15–25 ft. long by about 9 in. thick, bolted to a mill floor-beam, between the sheer-trees. Sometimes a cartwheel on the tail-pole supported the weight of the steps when they were raised; sometimes a 'link chain' reaching down from the top of the mill-body served the same purpose.

THE BRAKE-WHEEL AND BRAKE

On the wind-shaft behind the 'neck-wear' is mounted a large face-wheel or brake-wheel ((25) in Fig. 3). Old brake-wheels are of oak or other durable wood (Pl. 21 a), firmly fastened to the wind-shaft with wedges secured with thick dowel-pins, and having parallel pairs of arms elaborately housed and bolted into the circular frame of the brake-wheel.

This pattern is known as the 'clasp-arm' type, but it is not the earliest pattern. The clasp-arm wheel seems to have been introduced about 250 years ago, the first known illustration of it being in a Dutch book in 1728. Before this the brake-wheel had had diverging 'compass-arm' spokes mortised through the wind-shaft (as mentioned later), and latterly iron spoke-arm wheels were fitted.

The rim or tyre of a wooden brake-wheel ranges from 7 to 10 ft. in diameter, and is generally built of six segments, or felloes, scarfed, bolted, and dowel-pinned together.

The arms of the brake-wheel (often halved and reversed for strength) can either be housed into the rim, or they can be parted sufficiently to grip the rim fore and aft (as they do at Brill; but for simplicity this is not shown in Fig. 3). They should have considerable depth from back to front—from 10 to 15 in.—to give rigidity, as the strain and thrust from driving into the mill gears is very severe. (See the Wenhaston brake-wheel, Pl. 21 a, which had been transferred from a wooden to an iron shaft, and therefore had thick angle-blocks of oak made to fill the resulting gap around the smaller diameter of the iron shaft.)

The cogs of the brake-wheel are mortised through its felloes, with a nail piercing each of their projecting shanks, and sometimes with slivers of wood at the side, or there may be wedges between the shanks, without pins. At Brill the cogs are mostly of cherry-wood, with a few of beech-wood; but apple is often used. (Iron wheels with wooden cogs are called mortised iron wheels (Pl. 22*b*), and both iron and wooden wheels can have segments of iron cogs bolted to them. Occasionally there are two rows of cogs on the brake-wheel, the inner row being used to drive auxiliary machinery.)

Round the brake-wheel there is a brake. At Brill it is of the usual pattern: four long wooden shoes 7 or 8 in. broad and 4 in. thick (26), coupled together by iron plates, and forming an almost continuous ring. The shoes (usually of ash, elm or poplar) are held in position by chains (27) or rods, and they are applied to the wheel by the wooden brake-lever or staff.[1] Viewed from *inside* the mill, the brake lever (28) is on the right in anti-clockwise or 'right-hand' mills and on the left in clockwise (or 'left-hand') mills. The brake always tends to be pulled *on* by the motion of the brake-wheel; this is the reason the lever is set on the right or left according to the direction of rotation. And the weight of the brake-lever, which is 9 or 10 ft. long and 6 or 8 in. thick, keeps the brake *on* when unattended.

It follows that a 'tail-winded' mill (one, that is, which happens to be running away *backwards*) is not easily held by the brake. (A mill ought in any case to be kept into the wind whenever possible, to relieve the sails from strains for which they are not designed.) Experience seems to suggest that a brake-wheel of dimensions adequate to control a common-sailer cannot be housed in the average tower-mill cap, especially if the sails are long; and such tower-mills would be troublesome to stop, when neighbouring post-mills were still under full control.

The brake-lever is mortised into the front corner-post of a post-mill, and is coupled to the brake itself with a forged iron link; an iron pin near the end of the lever engages, when the brake is off, with an iron brake-catch, shaped like an inverted swan-neck and generally suspended from the head rail of the mill wall, or from the rafters of a mill cap. Thus the lever holds the top arc of the brake off the wheel, while the positioning chains keep the brake in alignment. A jerk on

[1] A continuous wrought-iron band brake 8 in. wide and ⅛ in. thick is sometimes furnished instead of wooden shoes. One or two mills in Norfolk and nearby had a 2 in. iron-rod brake running in a special circular iron groove mounted on the brake-wheel, and very effective it was; Patcham Mill, instead of the usual lever, had a screw control such as is more often found in France.

the brake rope, passing from the lever over a pulley on the head-rail and down to the ground (for use when reefing the sails, etc.), will dislodge the lever and apply the brake.

At Brill, holes were made in a wall-stud to take a rod or bolt for holding down the brake-lever; and these holes and the corner-post mortise may still be seen in the left-hand wall as well as the right, owing to the reversal of rotation over a century ago.

DRIVING GEARS

Except in head-and-tail mills (see Chapter 1), the brake-wheel engages with the 'wallower' or crown-wheel ((29) in Fig. 3) at the top of the upright spindle or main shaft (30). This wallower wheel is usually 4 in. thick if of iron, and 9 or 10 in. if of wood. The wallower (it 'wallows' about, and is really an upturned bevel with a slight resemblance to a royal crown) is one-half or one-third the size of the brake-wheel; if the wallower is big, the sails must run fast to produce the necessary millstone speed, and the 'angle of weather' of the sails is determined with this in mind. Iron wallower wheels are either keyed or wedged to their shaft while wooden wallowers are always wedged. In very old windmills the wallower is of solid wood, bound with iron.

Cross-tailed gudgeons are mortised into the top and bottom of a wooden upright shaft, the top one running in a sprattle-box (31) bolted to the spindle-beam, and the bottom one in a toe-brass (a square brass block cupped in the centre) contained in a bridging-box or centering block upon the main bridge-tree or lower spindle-beam (32).

The tail end of the lower spindle-beam in a post-mill, being in front of the main-post, has to be suspended from the face of the crown-tree (8) by a very strong and well-braced substructure; and all spindles of any size or importance are furnished with split 'brasses', or plain brass collars.

Wedged to the upright shaft is the great spur-wheel (33), which in this instance (Brill) is of old-fashioned wooden construction, being of clasp-arm type with twin sets of arms above one another to give lateral rigidity. This spur wheel drives the iron pinion 'stone-nuts' and spindles (34), the spindles being of iron. At Brill the stone-nuts were slid up and down into engagement (by means of a 'rigger' or yoked lever) on a splined and slightly tapered spindle. The yoke of the lever, which reached over the nuts, had two chains and hooks suspended, for gripping the arms of the nut. Sometimes a group of three or four cogs called 'slip-cogs' are made removable from the nut by taking out one or two iron pins or skewers, in which case the nut is fixed on its shaft. More often, the practice, almost universal in watermills, is adopted of lifting the nut on a screw-operated jack-ring instead of having a yoke lever; a slot in the eye of the nut engages with a long 'reed' or key which registers in a key-way in the spindle.

OVER-DRIVEN AND UNDER-DRIVEN STONES

The majority of surviving post-mills have the stones under-driven on the upper or 'stone' floor, but the hurst-frame (usually on the bottom floor) was much favoured in the south Midlands when reconstructing head-and-tail mills as spur gear mills (see Chapter 1, p. 12, 'Layout of Millstones'). Built like a massive bench, about breast high, it supported two pairs of stones side by side, and it usually had the driving gears beneath.

The stones[1] were most often one pair in Derbyshire Peak stone, fashioned from a single slab, for barley and grist milling; and one pair in French 'burr' for wheat, each about 4 ft. diameter, the runner-stone being 12 in. thick when new, and weighing over a ton, the bed-stone (which is fixed to the floor) being thinner and very often adapted from a worn 'runner'.

In head-and-tail mills, which are without intermediate gears, the millstones are necessarily 'overdriven', that is, from above; so are many spur-gear mills if there is head-room enough between brake-wheel and stone-floor, for this is more efficient on account of direct-ness of drive (and there is therefore less resistance from complicated mechanism). In a post-mill with little head-room, however, lack of

[1] The late Mr Cripps of Cuddington Mill—a ninety-year-old windmiller and watermiller—said there ought to be at least three pairs, especially with 'common' sails, so that you could 'ride out' a heavy gale and not have to stand by idle, which would do as much damage in a day as running would do in months.

space is a hindrance when dismantling the stones for re-dressing. But, against this, 'underdriven' gears are inaccessible in a cramped mill, and collect a great deal of dirt and flour-dust from above. Undoubtedly, the earliest windmills with one pair of stones were overdriven, and the underdrive arrangement originated in water-milling country, where the drive comes up from the waterwheel below.

POWER TRANSMISSION FROM SPINDLE TO MILLSTONE

We have seen that any corn-mill may be described as a machine for grinding grain into meal.

It may use for its motive power, human strength direct (as in the quern), animal power, water power (as the watermill), wind power (as the windmill), or steam power, oil, or electricity (as in modern mills).

The method used in the past for reducing grain to flour-meal has been to grind it between two flat circular stones, one of which revolves.

This essential mechanism for grinding, therefore, is the centre of any corn-mill, which must contain one stone that revolves upon another.

We have seen how wind power has been harnessed to provide a force that will drive such a mechanism, but we must now look to the stones themselves and learn how they are mounted in position, how one of them is made to revolve, how they are fed with whole grain, how they reduce whole grain to meal and deliver it as the outgoing product of the mill.

The mill sails turn and revolve the wind-shaft on which they are mounted. On the wind-shaft a great cog-wheel (called the brake-wheel) is fixed, and as the brake-wheel turns it transmits its revolving, through cogs, to an upright spindle upon which must be mounted (and thus revolved) one circular stone over another. Generally, the upper stone revolves above the fixed nether stone, and while it is revolving, grain, fed in at the millstone's eye or centre, eventually makes its way out between the opposed faces of the two grinding stones as grist, or flour-meal.

We have understood the principle of the mounting, in the top of the mill, of the inclined wind-shaft, and its pivoting so that the sails can revolve it. We have understood the gearing of the great driving wheel (the brake-wheel) and how it engages with the wallower, and thus with the main shaft, which is geared to a stone-spindle for each pair of stones. We therefore understand how a revolving or grinding motion is made available to the miller in his mill, at the stones, and we are ready to examine what is the most complicated portion of the mechanism depicted in our windmill diagram.

THE MILL-RHYND AND RUNNER-STONE

The revolving spindle is coupled to the upper stone of the grinding system by an iron block (the mace), which is mounted on the spindle and revolves with it, and clasps a stout short iron bar (the bridge) fixed across the eye of the upper stone, so that the stone will be carried round by the spindle (see Fig. 4).

To keep this stone just out of contact with the nether stone at all points (which means maintaining it in a balanced state) the iron bar or bridge ((35) in Fig. 3) is accurately centred in the stone (36) and is made to rest upon the apex of the spindle. The stone has two recesses into which the bridge is first wedged (across the eye) with wood (when assembling the stone) until the balance is found by trial and error. Then the interstices round the wedges are filled with molten lead; the stone is then turned up the other way (see Pl. 28 a).

The mace or mill-rhynd (37) is located on a squared boss on the spindle; if the mace has forked ends or jaws, they register with a pair of shoulders formed on the underside of the arched bridge-piece, without disturbing the self-balancing of the stone.[1]

---

[1] Should a new mill-rhynd or stone-spindle be found necessary, the eye of the rhynd should be machined out if need be to enable it to set lower down on the shoulder of the spindle.

In no circumstances should the recesses in a French burr-stone be cut farther into the stone as an alternative, or you may cut through the burr into the plaster-of-Paris backing that holds a burr-stone together, and it will be ruined. On the other hand, as, in the course of re-dressings, the face of the stone is chiselled away, the shoulders of the cross-bar will stand proud, on the underside, so they must be made to recede, within reason, by judiciously gouging out the recesses in the stone with a cold chisel, and re-leading the bar. Thus there is no margin left for resorting to a similar process for other adjustments.

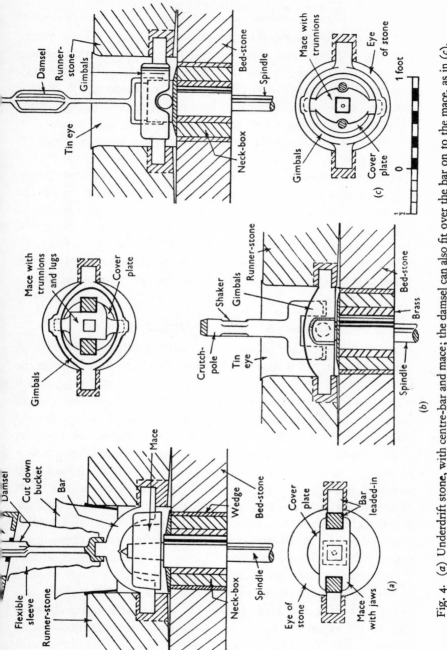

Fig. 4. (a) Underdrift stone, with centre-bar and mace; the damsel can also fit over the bar on to the mace, as in (c). (b) Overdrift stone, with gimbals. (c) Underdrift stone, with gimbals.

47

A less satisfactory coupling procedure is to have a mace provided with trunnions (instead of forked ends) on which rest two recesses formed in the underside of a stout iron ring or 'gimbals' (Pl. 22 b). The other two sides of the gimbals have trunnions which register with iron or bronze liners leaded into the recesses in the eye of the stone, thus supporting it, instead of having a rigid bridge or cross-bar.

When the stones are driven from below, the mace or bridge, as the case may be, will have provision made for receiving the foot of the damsel (see below) so that it revolves with the millstone. Referring again to Fig. 4 it will be seen that an overdriven stone has the foot of the driving shaft forked, so that it fits down over the mace, which has a shoulder formed on each side, against which the legs of the fork are made to bear. This forked shaft is called a 'crutch-pole' or 'quant', because it oscillates like a quant-pole when freed from its upper bearing (the glut-box) in order to disengage the stone-nut.

BALANCING THE RUNNER-STONE

A runner-stone cannot balance unless the stone-spindle which supports it be true, and this in olden times was achieved by locating the foot of the spindle in a big wooden block about a foot long, which rested in a rebated portion of the bridge-tree, with a space for wedges on either side of the block so that it might be wedged in any desired exact position. This device survives in the derelict post-mill at Normanton Common (see Pl. 16 d), in conjunction with an obsolete form of governor, but was referred to by John Sutcliffe in 1816 as an outdated arrangement. He advocated (and incorrectly implied that he originated about the year 1785) the practice of setting the spindle in a bridging box some 4 in. square (Pl. 15 b) which in turn accommodated a 2½ in. square brass step centred by four grub-screws in the walls of the box, for the purpose of setting the spindle perpendicular.

Provision is made for inserting several variable balance weights of lead in the back of the upper wheat stone (which consists of a group of blocks cemented together) to allow for its inequalities in texture; but as the 'pivotal plane' (about which a revolving millstone balances itself) is located midway through its thickness, the 'standing balance' of the stone at rest may be different from the 'running balance' when in motion. Thus the heaviest balance weight may then press down in an effort to coincide with the pivotal plane, as the stone gathers speed, and so throw it askew. Therefore a 'false hoop' should be fitted round the lower periphery of the stone, and a counter-weight of lead inserted in it, below this balance weight. This is less likely to be necessary with the adjustable weights mentioned in Chapter 1, p. 14.

THE GRINDING PROCESS

Space is left for grain to pass around the rhynd or mace, which occupies only a portion of the 10 in. diameter centre 'eye' or 'waist' of the runner-stone; and the eye has a metal lining (the 'tin eye') in a French stone, or a 'drag-stick' of hazel fitted across it in a Peak stone, to prevent the grain from clogging.

The meal discharges round the skirt or periphery, between the two stones, after having been ground by the radiating furrows and the fine 'featherings' or 'stitching' on the 'lands' between the fur-rows, on the corresponding faces of the two stones.

These grooves, or 'dressings', on the stone are set at such an angle that when the stones are in motion face to face the furrows keep crossing each other like a succession of scissor blades; and the broken grain is driven outward by this action of the grooves, and by centrifugal motion imparted by the upper millstone swirling it round.

Periodically the furrows are re-dressed by the miller (see Chapter 7, p. 101), or by a professional stone-dresser, with a highly and specially tempered steel mill-bill set in a handle called a thrift. He also uses a heavy bill, sometimes termed a 'pritchell' which resembles a cold chisel sharpened to a point, for the preliminary deepening of the furrows. An exceptionally skilled 'dresser' can cut sixteen featherings to an inch, but eight or ten feathers are more customary. Dressing, and subsequent re-balancing, of the stones is a fine art, on which much of the efficiency of the mill depends; indeed, it is claimed that the work should be so finished that the 'nip' or closeness of the stones will only permit of a piece of brown paper being placed be-tween them at the centre, and a piece of tissue paper at the periphery.

THE BED-STONE

The lower stone or bed-stone (sometimes called the nether stone), is set 4 or 5 in. into the floor, where it rests upon the stone-beams, which are built into the main floor-beams to form a square frame of 6 in. timbers, some 2½ ft. across; the stone is firmly wedged round the sides, and is then closely encircled by a wooden kerbing fixed to the floor-boards. The hole or 'eye' of the stone, accommodating the spindle, is usually square, and has an iron or wooden bush or neck-box to exclude the grain; for lubrication this can be packed with binder twine dipped in melted mutton fat or candle grease. Iron boxes may have three or four brass inserts adjustable with metal wedges that are tightened with screws from beneath. To the side of the runner-stone there is nowadays affixed a brush or metal 'paddle' to augment the action of draught and friction in sweeping the meal round the annular space between stones and stone-case or vat so that it passes through an orifice or spout to the lower floor, otherwise some meal is lost at each grinding.

The exit orifice, cut through the curbing of the bed-stone, is 4 or 5 in. square, although the orifice has occasionally taken the form of an annular slot round the stone, which seems to offer no advantage. The top of the stone case, and the feed hopper and spout (described on p. 53), are seen in Pl. 16b.

As a bed-stone wears down, it is levelled up with slabs of wood, and smaller wedges, laid on the stone-beams and framing as required, so as to keep the working face at the original height. Eventually it will be replaced, as often as not by a worn runner-stone; and a new stone will be installed as a runner, for which purpose a good weight and thickness are required. Although not absolutely essential, it is very desirable for both stones to be of the same diameter.

Stone-cases, 'hoopes', or vats, enclose the stone whilst at work to keep the meal and dust under control; and these are halved and clipped together if there should be insufficient clearance beneath the head-wheel for lifting the stone-cases over the stones.

TENTERING GEAR

The stone-spindle is seated on a brass or gun-metal footstep or cup, centred in a cast-iron socket or pot,[1] which rests on the wooden lighter-bar or bridge-tree ((38) in Fig. 3), and the tail end of the latter is loosely mortised into a convenient post whilst the front end rests, in this particular mill, upon a cross-bridge-tree or 'brayer' (39). This again is mortised at its inner end, and coupled at the other to an upright iron rod (40) with a hand-screw or socket-spanner adjustment (41) for regulating the fineness of the meal.

At its upper end this rod is coupled, through a link, to the steel-yard (42), the yoke of which engages with the collar of a centrifugal governor (43); and the latter, which serves both pairs of stones in Brill Mill, is belt-driven, with wooden drums, from the upright shaft on the stone-floor. This, however, is a rare arrangement; the governors are seldom elsewhere than *beneath* this floor as seen in Fig. 5, and separate governors are often provided for each pair of stones. Cross-bridge-trees (sometimes called 'cloves' in the Midlands) are often dispensed with, and mortising may be for single or double tenons, the latter giving a broader bearing surface. The tenons, and the shoulders of the bridge-trees, should be slightly rounded in longitudinal direction, on their undersides, so that they ride smoothly on their seatings while the mortises should be deep, to allow for raising or lowering of the bridge-tree from time to time as the stones wear down.

John Sutcliffe suggested in 1816 that the bridge-tree, being pivoted one end and raised and lowered the other, did not keep the stone-spindle on a sufficiently even keel; therefore, he said, the bridge should remain stationary, and the spindle should have a screw thread on it and a pair of bevel wheels to be operated by a cranked handle, much the same as the spindle is raised on a modern motor-car jack.

Sutcliffe also contended that a fault common to all classes of mill-stone mill was that the heating-up and expansion of the stone-spindle while at work must cause the runner-stone to be gradually raised,

---

[1] Another satisfactory arrangement is to have a brass sleeve or collar, and beneath the spindle a *penny*, which will last a long time.

and that this could, if not attended to, cause a variation in the quality of the flour. Some millers, he suggested, did not notice this effect and, having adjusted the stones a bit during the day in an effort to secure the best results, started up again next morning without realizing that the contraction of the spindle (due to cooling down) had

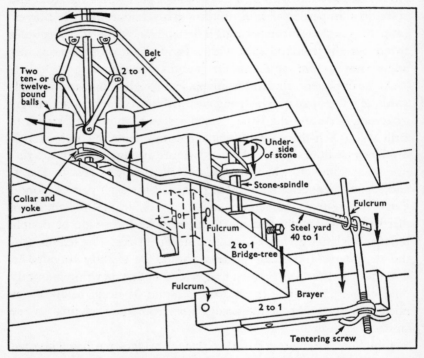

Fig. 5. Tentering gear controlling the tail-stones of Syleham Mill; the heavy arrows indicate direction of movement when the governors accelerate.

caused the stone to settle. They therefore made no adjustment until perhaps half an hour or so later, when they had observed the difference in the grinding compared to the grinding obtained overnight. He knew of no automatic remedy, he said; the miller must note and understand the cause of the variations.

Quite often, the tentering gear will be different for different pairs of stones, and sometimes the layout is particularly complicated.

On gathering speed, the governor balls fly outwards, causing the collar to raise the steelyard (Fig. 5); this, acting through the bridge-trees, *lowers* the upper stone (which is tending to ride up on the grain as it feeds in at an increasing rate), and ensures thorough grinding at all speeds. But note that the governors act according to momentary variations in the force of the wind, while the miller must make his own general adjustment by hand-screw, to suit the kind of grain and the average force of wind on the day.

Should the driving belt break, the governor balls will drop and let the stone *up*, not down, thus obviating risk of fire from friction. It should be noted that the stones are slightly apart when stationary even though not supported by a layer of grain, nor held up by the hand-screw or 'lift-tenter'. This can be proved by lifting the governor balls with the hands, when the bridge-tree and stone-spindle will be seen to drop slightly, bringing the stone with them of course, to rise again when the balls are lowered. This happens through the multiple leverage in the irons and bridges—thus the two balls, weighing some 10 lb. apiece, usually have at least a 2 to 1 leverage in their own arms; the steelyard, owing to the position of the pivotal link, has a 30 or 40 to 1 leverage; the brayer 2 to 1 (because the bridge-tree rests midway along it) and the bridge-tree a further 2 to 1—thus the balls hold up the 25 cwt. stone by a leverage of some 200 or 300 to 1.

As a millstone loses perhaps one-third of its weight before being discarded, provision is made for varying this leverage by engaging another of the several notches on the edge of the steelyard with the pivotal link.

THE GRAIN FEED

Grain stored in bins or garners ((44) in Fig. 3), on the top or bin-floor, is fed down to the stone-floor by a spout (45) into the hopper (47) which is supported by the wooden 'horse' (46); and from this it passes down the feed shoe or 'slipper' (49), which in turn is pivoted on a peg at one end and held against an iron damsel (48)—an extension of the stone-spindle—by a cord. This cord is attached to the end of an ash or hazel rod (50) which is nailed to the top or the inside of the stone-case (Pl. 16b) so that it acts as a spring. The damsel supersedes an even more archaic contrivance called the 'clapper' (which was mounted on top of the millstone to make it strike the feed shoe). As the damsel revolves, it vibrates the shoe with a loud, rhythmic clatter (hence its name!), thus shaking down the grain into the stone's eye; another cord (51) is carried over guide-pulleys to a thumbscrew (52) on the lower floor, by which the miller, testing the fresh meal in the sack, can raise or lower the shoe to vary the feed. He may also adjust a 'spattle' on the side of the hopper, where it feeds into the shoe; but the hopper cannot overflow, because the chute extends down into it. (In the northern counties there is no hopper; the chute feeds into a big elongated shoe.)

In some mills, especially when gristing, the sack is hung on three hooks on a suspended wooden bar called a 'sack-boy'; but this was often impracticable in hurst-frame mills. At Brill the meal passed down a wooden chute (53) into a bin, which entails extra work, but enables the meal to be mixed with the miller's wooden shovel or barn scoop before filling the sacks. This ensures more even texture if the wind force or the quality of grain varies appreciably. It also cools the meal more quickly, ready for the flour dresser.

To inform the miller when the hopper needs replenishment, a bell-alarm or 'warbler' is provided; it is a piece of leather placed at the bottom of the hopper before admitting the grain, with a cord leading over pulleys to a suspended bell (54) so placed that when the leather is released the bell-bar falls against a moving part (such as the damsel or the cogs of the great spur-wheel) making a continuous alarm, audible from the mill yard. Sometimes the vibration in the mill causes the bell to emit a pleasant musical note all the time, above the general din of the machinery; but some millers bury the bell itself in the grain, and when the hopper is nearly empty the bell rings because of the movement of the grain.

SACK TACKLE

An important item in all windmills and watermills is the sack hoist; and in the Buckinghamshire post-mills a wooden drum ((55) in Fig. 3) is wedged to the wind-shaft just behind the brake-wheel (although elsewhere it is almost invariably ahead of the wheel), and a driving chain passes thence over a smaller drum on the chain barrel or bollard (56) which is normally of wood with iron longitudinal strakes to take the wear. Only occasionally is the drive by chain— another isolated example being at Icklesham Mill, Sussex—for a belt or 'wrapping connector' is almost universal. The bollard, which is furnished with gudgeon pins, is engaged by raising the driven end with the lever and rope (57) at Brill, but in post-mills elsewhere the lever is indirectly operated by a rope and jockey pulley coupled to one or more secondary or purchase shafts. Brill Mill originally had an even rarer arrangement than the chain drive; for the wind-shaft and bollard each had cogs mortised or housed into them, giving a direct spur drive (as still found in Chinnor Mill nearby), but this could only raise the sacks—they had to be lowered by hand, whereas they can normally be lowered by 'slipping' the belt drive. Another arrangement, often met with in Norfolk and Sussex, is a spur-gear meshing with the inside faces of the brake-wheel cogs, and thence by belt.

Each sack is gripped about its neck by the chain (58), the end of which is looped upon itself, and the gear put in motion, when it passes up through the sack traps or hatches (59) to the top floor. An assistant unhooks it and returns the chain through the central holes in the traps; the trap doors, which have leather hinges, automatically fall shut as the sack goes up, thus safeguarding those walking on the floor.

The sack bollard, or an intermediate control shaft, can alternatively be friction-driven from the outside of the tail-wheel rim as at Outwood, or the inside rim or sole of the brake-wheel as at Dale Abbey, Derbyshire, and elsewhere.

FLOUR DRESSING

Wheatmeal—that is, finely ground wheat which would make whole-meal bread—was usually returned to a hopper on the top floor and was thence passed through the dresser (60), having first stood by for several days to cool off, when it would separate better. By the dressing process the bran particles were removed, and it became known as millstone flour, as distinct from wheatmeal.

Usually a cross-shaft passing over or under the wind-shaft, and carrying an iron skew gear, conveys the power from brake-wheel to dresser-spindle in a post-mill, engagement being made by a fulcrum lever and yoke arm while the machinery is stationary; but at Brill a wooden pinion (61) is mounted on a short shaft in a swinging frame (62). A belt, which tightens when engaging the pinion, leads to the tail of the mill, where a cherry-wood cog on a short counter-shaft (63) is geared to the iron spindle of the dressing machine. In the big inclined case which straddles the middle and lower floors is a wire cylinder (64) 5 ft. by 18 in., supported by numerous wooden rings, and provided with finer gauze near the top than at the lower end, so that the meal entering at the upper end is separated into firsts, seconds and 'supers'—that is, first and second quality flour; and 'supers' or middlings for the better quality animal foods—as it encounters the parallel revolving brushes (65) mounted on the dresser spindle. Such material as is too coarse to pass through at all emerges (at (66)) in the form of bran. The dresser preserves the nutritious 'germ'—the bud of the wheat berry; the bran goes to livestock, horses needing a 'pick-me-up', or rabbits; or it can be manufactured into proprietary foods for the breakfast table.

Various grades of flour can be separated from bran by passing the meal through a bolting machine, which revolves at only about one-tenth the speed of a dresser, and turns the meal over and over as it passes through; the meal cools off in the bolter and can therefore be dealt with straight from the millstones instead of standing by to cool, and the flour can also be bagged off as it comes from the bolter.

The wooden spindle of the bolter, driven by belts and bevel gears (the same as a dresser), has several sets of radiating arms, which carry six longitudinal bars fixed to circular end-plates. Over these a seam-less bolting cloth is stretched like a long sleeve (about 6 ft. by 20 in. in a windmill, and larger in a steam-mill). The original sleeves were of wool, but later of calico, and now of silk; and they are attached to leather collars at both ends, with cords for fixing them to the reel.

Meal is fed into the upper end, with a leather flap or guide resting in the mouth of the cylinder, and the revolving sleeve billows out a little under the weight of the flour and rubs against several fixed longitudinal bars, thus assisting the flour to pass through the 'silk'; and the bran is discharged at the end, as in a dresser. Sometimes the miller himself darned the bolting cloths; and one told me that, unwisely using a candle to look for holes in the cloth, he was greeted by an explosion from suddenly ignited flour dust, but was uninjured.

Various forms of 'centrifugal' bolters were introduced later for removing the remnants of flour adhering to bran, pollards and sharps, the meal being driven through by a high-speed centrifugal blower against the walls of a vertical bolting sleeve. One old miller told me that when this machine was installed (usually in watermills or steam-mills) they were at once able to command a better price for their flour. James Bell in 1850 said that his patent 'centrifugal' made 800 revolutions per minute.

When arranging the auxiliary drive for dressers and bolters in a post-mill, the reason for employing a skew-gear with spindle right across the mill and the long driving belt at the other side, out of the way of the brake-lever, is doubtless to provide a more rigid 'take off' from the brake-wheel. But from the extended spindle it is also convenient to have another belt driving other machines, such as a separator for cleaning dust or sand from the grain, an oat crusher or roller, a chaff-cutter, grindstone, a smutter for cleaning 'smutty' wheat, an eccentrically driven joggling-screen or 'scrier' for dressing corn, screw- or worm-conveyors, and so on.

Groats were made, in the north and in Wales (and more often in watermills than windmills), by roasting oats in a kiln, grinding them in millstones less finely adjusted than usual, and putting them through the groat machine, from which the dust and hulls or husks were blown into a 'husk' cupboard by a winnowing machine.

Portable equipment generally included a sack-truck, large mill scales—sometimes very old and partly of wooden construction—and a couple of big triangular wooden blocks, with inserted handles like a short broomstick, for slipping under the stones as they were levered over for dressing.

Chapter IV

SAILS AND MILLSTONES

✣

'*SPRING*' *AND* '*PATENT*' *SAILS*

IN addition to 'common' or cloth-covered sails, already described, 'spring' sails and 'patent' sails were much used.

The framework of spring sails consists of eight or ten sail bars, spaced 3 ft. apart and forming a series of bays, in each of which are three shutters 4 or 5 ft. by 11 in., wood or metal framed, and covered by wood or canvas. A hemlath is of course provided, this and the sail-bars being of about $2\frac{1}{2} \times 1\frac{1}{2}$ in. deal timber secured with coach-bolts: and backstays are provided for most of the sail-bars. A leading wind-board is fitted unless the sail is 'double-shuttered', when there is a second set of shutters 2 or 3 ft. long on the leading side, making the sail 7 or 8 ft. broad overall; and the last stay-bar can in this case be carried straight across the front of the whip if desired (Pl. 31 *a*). In the Fylde and the Wirral Peninsula every shutter had its own sail-bar and bay.

Each shutter is pivoted by means of fulcrum pins seated in iron 'thimbles', and it is coupled by an iron lever to a wooden sail-rod, $1\frac{1}{2}$ in. square in section, which thus operates a complete set of shutters simultaneously. The shutters can open partially to 'spill the wind' and relieve the strain on the sail in a sudden gust, and the spring, which can be elliptical, semi-elliptic, or coil-type, will close them (Pl. 24). But such sails need to be stopped by the brake for reefing by means of a lever and rod on each sail; and some smock-mills and tower-mills with common or spring sails and no stage had a portable reefing stage like a large table on wheels, which was trundled round outside the mill.

'Patent' sails incorporate a notable improvement in control, enabling the miller to reef and adjust them whilst in motion.

Referring to Pl. 25 it will be seen that the sail-rods are connected to a 'fork-iron' which is coupled to an iron 'triangle' pivoted near the hub of the sails. Each triangle is connected to the central cross or 'spider' by a bridle iron. The spider is fixed to the end of the striking rod, which passes right through a 2 in. bore along the wind-shaft, and is coupled to a sliding bar or rack by a ball-and-socket joint. The rack is geared to a spindle and 'purchase wheel' which carries a chain on which the miller hangs the weights to close all the shutters simultaneously. In Lincolnshire and Kent the weights hang from a long pivoted iron bar; and in Norfolk and Sussex a long pole suspended from the fanstage steadies the chain.

Some sixty years ago Charles Edwin Hammond of Clayton Mills, Brighton, attempted to regulate 'patent' sails with a very large pair of governors and two clutches.

MITRED SAIL-SHAFTS AND MULTI-SAILED MILLS

When a central iron cross is employed on the head of the wind-shaft to carry the sails, as at Coleby Heath (Pl. 35), and most other Lincolnshire mills, instead of a poll-end or canister, the sail-shaft becomes in effect a combined whip and stock. All the shafts, which are then called 'backs', are mitred together at the centre, and each is attached to its iron arm with three or four big U-bolts. At the heel each 'back' is about a foot deep and 9 in. broad, tapering to some $4\frac{1}{2}$ in. square; and the advantage of this arrangement is that the damage arising from wet lodging in the canister is obviated. An average 'patent' sail of this description probably weighs a ton, and it eventually breaks, if not renewed in due course, at the outer U-bolt.

One advantage of the 'cross', and possibly the reason for its introduction, is that it enables more than four sails to be fitted to a wind-shaft, which is almost impossible with a canister-head.[1] From experience in Lincolnshire particularly, there is little doubt that the ideal is five or six sails, and if six, they should be single-sided, to give

---

[1] Ashcombe post-mill, Lewes, with a triple-headed canister, is the only known exception.

the wind a chance to escape from each sail, so that the sail following does not have too many eddies to contend with.

There appear to have been about sixty or seventy multi-sail mills built in England, of which half were in Lincolnshire. The present writer has recorded the following examples:

Eight-sailed tower-mills were at one time to be seen at Heckington, Market Rasen and Holbeach (Lincolnshire); Wisbech (Cambridgeshire); Eye (Northamptonshire); Much Hadham (Hertfordshire); and possibly Old Buckenham (Norfolk), which is now a four-sailer.

At least eighteen six-sailed towers were built in Lincolnshire; four in Norfolk; two in Cambridgeshire; a pair together in Derbyshire, and one separate; two in Nottinghamshire; one each at Wymondham (Leicester), Hessle (Yorkshire), Nowton (Bury St Edmunds) and Leighton Buzzard ; six-sailed smocks at Great Chart (Kent), Chailey (Sussex) and Hounslow Heath ; and a post-mill at Lewes.

Five-sailed towers stood at York and Seaton Ross; eleven were in Lincolnshire; one converted from a four-sailer at Hartshill (Warwickshire); and five-sailed smocks stood at Newcastle, and Sandhurst (Kent).

Two interesting variations of the adjustable shutter principle have occasionally been employed in England; they are the 'roller-blind' sail and the 'annular' sail or 'wind-wheel'. The former, invented by Captain Stephen Hooper, achieved a little popularity in Yorkshire, where it survived on Tollerton Mill and Seaton Ross Mill. Every shutter took the form of a spring roller-blind, all being coupled together by webbing straps; and the twin control rods, forming a large V on the sails, were called 'air-poles'. This device was not very durable.

Annular sails, with 'patent' controlled shutters arranged in a 40 or 50 ft. circle on an eight-armed wheel or frame, were employed in East Anglia on Haverhill tower-mill (Pl. 27a), Boxford smock-mill and Feltwell and Roxwell post-mills (all demolished), and the wheel was powerful and efficient, but expensive to maintain. Catchpole's patent auxiliary shutters (see Chapter 1, p. 9, 'Sail Developments') were fitted to Buxhall tower-mill, and Gedding and Wetherden post-mills, all in Suffolk, where they were popularly called 'sky-scrapers'.

ANGLE OF WEATHER

James Ferguson (*Lectures on Select Subjects in Mechanics, etc.* 1776) asserted that sails ought to make an angle of 54¾ degrees with a plane perpendicular to the axis on which the arms are 'fixt'; but the vane should be given a twist, and the ribs should decrease in length, giving the vane (i.e. the sail) a curvilinear form, so that no part of the force of any one rib be spent upon the rest, but all move on independently of each other; 'and we see both these features exemplified in the wings of birds'. (Tapered sails such as Ferguson described are still found in Holland, Anglesey and elsewhere, but modern sails everywhere have a twist, warp, or 'weather', to enable the wind to exert a similar pressure on the tip, travelling at 30 m.p.h., as it does on the heel, which is moving slowly.)

No mention was made by Ferguson of the primitive practice now found only in Somerset and backward areas abroad, of setting the 'axis' (i.e. the wind-shaft) in a *horizontal* plane instead of inclining it downwards to the tail. All competent millwrights agree that the sails should be 'so set back as to bring the top of the uppermost sail well within the diameter of the plane of the roof, throwing part of the weight well back, so diminishing the whole forward pull of the sails. Then the steps and tail-beam on the side farthest from the sails form a nearly equivalent counterpoise which both steadies the structure and satisfies the eye' (Gertrude Jekyll, *Old English Household Life*).

It is also desirable, in the present writer's opinion, to set the individual sails forward a little from the centre (especially at the tips, as he has done at Brill), so that they revolve in a plane resembling a shallow saucer, thus 'gathering up the wind' to best advantage; but it is necessary to consider how to shed the wind as each sail descends, otherwise it will 'lodge' in the sails like excessive tail-water in a water-wheel, and will act as a drag upon them. As each sail passes the base of its orbit, it is supposed to pick up the wind again with a 'swish'; according to some millwrights this is the test of a correctly designed sail. Sam Clarke in Suffolk and an old Norfolk millwright, Fred Goffin, were reputed to make good sails. The 'dummy' sails on Sproxton Mill, Leicestershire, were set forward from the centre.

ROTATION OF SAILS

Most early post-mills had but a single pair of stones driven off the brake-wheel; but those surviving into the nineteenth century, including Brill Mill at the outset, had a second driving wheel mounted near the tail of the wind-shaft, the brake-wheel in this case being called the head-wheel and the other the tail-wheel, and each was geared directly to a millstone. The wind-shaft at Brill is mortised to receive the arms of the two 'compass-arm' wheels, i.e. wheels with four arms, like the points of a compass, passed right through the shaft; but the tail-wheel and tail-stones constitute a serious obstruction and danger when a flour-dresser is installed, as the tail is about the only convenient place for this contrivance; and Brill Mill and most of its neighbours were refitted with the stones side-by-side in the head, driven by a new brake-wheel, with the usual intermediate spur-gear.

She was at the same time *reversed from clockwise to anti-clockwise motion of the sails* to avoid reversing the rotation of the stones, because the oblique 'dressing' on their grinding faces would otherwise have to be inclined at the opposite angle.

'Clockwise' stones, revolving 'with the sun', are easier to dress, because the stone-dresser sits on the right-hand side of the stone, as it were, with his left elbow steadied on a sack of straw on the stone, while using the tool in his right hand (Pl. 28 c). An anti-clock stone, running 'against the sun', can only be dressed by sitting on the left-hand side, with the working arm over the stone and left arm unsupported. Consequently he tends to slip off the stone, since he has no purchase over the job. The very earliest millstones were merely 'pock-marked' (as practised in primitive countries even today). This rendered the direction of rotation immaterial, but one imagines the sails were instinctively made 'clockwise' even then.

WHEEL-AND-CHAIN LUFFING GEAR

Post-mills were not the only mills luffed into the wind by a tail-pole; a corresponding pole, reaching down from the mill-cap, was used on smock-mills and tower-mills for many years; and, indeed, this remains the standard method of winding such windmills in Holland and elsewhere today. English and Dutch mills have this pole strengthened by a set of outriggers extended from the flanks and tail of the cap, and the pole generally carries a winch (which may be geared to increase the leverage) for use with chain-posts placed round the mill plot. An internal winch is embodied in the cap of Chesterton tower-mill, Warwickshire. Another rare arrangement, occasionally used in England but more often abroad, was a capstan wheel mounted within the cap and geared to a cog-ring on the mill-curb; and this may have been the forerunner of the wheel-and-chain device favoured in the Fylde and Anglesey, etc., where the wheel is outside the cap, sometimes beneath an extended hood. Constructed like a cart-wheel it may be 5–10 ft. in diameter, with a V-groove or a series of iron forks or Y's projecting outwards from its rim to accommodate an endless control chain; suitable intermediate gears, sometimes incorporating two or three counter-shafts side by side, are provided within the cap.

FANTAILS

Fantails were eventually applied to most East Anglian post-mills, where they were generally carried on a 'fly-frame' over the steps (Pl. 1); but one or two Norfolk millwrights adopted the Sussex method of mounting them upon a triangulated timber frame attached to the tail-pole and steadied with iron ties, and a Hagworthingham (Lincolnshire) man fitted a similar device at Chinnor. Why the Norfolk mills should have some affinity with those of Sussex rather than Suffolk, I cannot say.

The tail-pole fantail is a poor arrangement because it imposes excessive strain on the tail-pole, and as it stands far back from the main structure it tends to swing the mill from side to side by acting more like a wind-vane than a fantail. The method standardized in Suffolk is to bolt a massive travelling frame and 'fly-posts' (about 20 ft. high and 6 in. square) on to the foot of the steps (dispensing with the tailpole), and secure the posts to the upper ends of the step timbers with two or three 'fly-strings'—horizontal rails of

4 × 3 in. timber—each side. These should be long enough to permit the fly-posts to slant back away from the mill, so that the inward pressure at the foot will counterbalance the outward thrust of the steps.

On East Anglian post-mills the fan or 'fly', from 9 to 12 ft. in diameter, drives iron tram-wheels on the travelling frame by means of iron mitre gears and shafting.

A fantail can be disconnected by unscrewing a coupling in the shafting; the mill can then be turned by operating the gears with a crank handle if the fan is damaged; or the mill may be cranked into the wind quickly if, as occasionally happens, it suddenly becomes tail-winded in a thunderstorm.

Owing to the high ratio of the multiplying gears (about 2000 to 1) a fantail turns the mill (or the cap of a tower-mill or smock-mill) only very slowly; and will not turn it at all if the wind is dead on the back of the mill.

According to one miller a sudden change of wind in a squall would push his fantail and steps a couple of yards round the tramway with a violent wrench, and another declared that he had seen his steps, with the 'fly' and tram-gear, lift a foot off the ground and come down with a crash when a heavy gust got behind it in a 'choppy' wind. Very rarely a post-mill has had twin fantails side-by-side, so as to catch a light breeze more readily.

In a steady changing wind a smock-mill fantail was recently noted to turn the cap through an angle of 180 degrees in 20 min.[1]

In construction the fantail, or flyer wheel, is almost standardized, with 4 in. square wooden arms or fly-stocks fitted into the six or eight sockets of a 'star wheel' or 'fly-star'. The segments or vanes (of matchboarding, battened on the edges) have their tips linked by iron stays. The sockets are open down one side to avoid lodging the wet. Flat rings of iron help to steady the arms, and it is essential to leave an 'open' centre to the fly, and not crowd the vanes into the

[1] Fantails have sometimes been transferred from a disused mill to another one; but as the top of the fantail on a smock-mill or tower-mill catches the wind more than the lower part, the top vane should be in line with the slip-stream from the sails (which deflect the wind to one side). If the top vane is set across the slip-stream, it will continually drive the mill a few degrees out of the wind.

middle, because of the importance, as mentioned in respect of the mill sails, of *getting the wind out of the fly*. Several Sussex windmills had a five-vaned fantail; Kingsdown (Kent) and Nounsley (Essex) had seven; Sutton (Norfolk) ten; and Sherburn-in-Elmet, Leeds, twelve.

THE FANSTAGE

On smock-mills and tower-mills only the cap and sails turn to meet the wind, the rest of the mill stays still. The post-mill is the only building that moves bodily with all its machinery. Beneath the fantail there is always a little 'storm-door' in the cap, leading on to a fan-stage from which the fan and the cog-circle on the curbing can be oiled and repaired. It is desirable to cast the cog ring in easily replaceable sections, for nothing is more likely to lead to serious damage than failure of the fantail to keep the mill winded owing to a missing cog.

As a fantail weighs up to half a ton, the outriggers, which are bolted to the main-sheers of the cap, may be made from a disused sail-stock, and the fly-posts are half housed and bolted to these, or they may stand in iron shoes weighing over half a hundredweight each.

From fantail to smock-mill or tower-mill curb the drive is conveyed in a variety of ways, a segmental iron cog-ring being bolted to the curb; and the transmission may be by spur pinion, worm, or bevel pinion. Some old mills had a cog-rail of large wooden teeth mortised into the outer face of the mill-curb, sometimes engaging with a worm constructed of wood, the worm being 6 or 7 ft. long and 10 or 12 in. in diameter. South Ockendon Mill, Essex, has this arrangement together with an intermediate iron worm-gear.

INTERNAL EQUIPMENT OF SMOCK-MILLS AND TOWER-MILLS

Interior machinery is much the same in smock-mills and tower-mills as in post-mills, but 'spoke-arm' brake-wheels with radiating arms are generally favoured. Should the radial arms be of wood, they are gripped between two big cast-iron plates or flanges, back and front, bolted together to form a hub. Often the spokes are of iron, cast integral with the hub, but this iron 'centre' is often bolted to a wooden rim. Iron wheels are either cast whole, or 'split' in halves and bolted together. Cast iron wind-shafts, with integral poll-end or a cross keyed on, are also popular in smock-mills and tower-mills.

The greater height of a big tower-mill provides for a better layout on the various floors, and the main upright shaft assumes more importance and larger proportions, being from 12 to 18 in. thick if of wood, and 5 or 6 in. if of iron; indeed, there is an 11 in. diameter hollow iron upright shaft built 32 ft. long, in sections, in Bramfield tower-mill, Suffolk.

A large gudgeon pin secures the shaft to the sprattle-beam (the main cross-member of the cap) and its foot rests in a large footbrass and pot on a massive bridge-beam having double tongues at either end engaging with two upright posts, so that the beam can be kept on an even keel if raised or lowered.

Sometimes the shaft is made half of wood and half of iron.

At the top of the shaft is the wallower, which may be solid wood in the older mills and cast iron in the later ones; and from the underside of the wallower a bevelled (or sometimes even a cylindrical) friction drum, on a horizontal sack-bollard or intermediate spindle, was brought into action by lifting the spindle with a lever or beam. This enabled a full sack to be gently lowered, as well as raised up from below, since the drum could be allowed to slip, as on a belt-drive, by easing the tension on the hand rope.

Iron wallowers not infrequently had a wooden under-face bolted on to provide a friction-drive surface; otherwise the sack-tackle would be belt-driven from a layshaft on the next floor down.

Besides the wallower and great spur-wheel, the upright shaft carries a crown-wheel geared to one or more layshafts, for driving the auxiliary machines, etc. There cannot be a 'take-off' from the brake-wheel as in post-mills, because the wheel is conveyed round with the cap whilst the machines remain stationary. An auxiliary cog-ring or a wooden belt-drum can also be bolted beneath the spur-wheel if required.

The stones can be overdriven if desired (i.e. gearing above the stones instead of beneath). Three or four pairs of stones are often provided in tower-mills, or even in smock-mills; indeed, Upminster smock-mill (Essex) has four pairs of 4 ft. stones disposed around a 9 ft. 6 in. spur-wheel, with 1 ft. 9 in. stone-nuts, and occupying only two-thirds of the circle, so that no less than six pairs could have been accommodated on the floor if required. The great spur wheel at Old Buckenham Mill, Norfolk, which is one of the outstanding tower-mills of England, is 13 ft. diameter, cast in twelve segments, and is possibly the largest wheel in any English corn windmill.

Three pairs of stones are controlled by one set of governors in some tower-mills by means of an elaborate system of steelyards, where the master steelyard regulates the wheatstones, and the others the grist stones.

The lower half of a tower-mill should afford plenty of room for filling and storing sacks, and a loading floor at 'cart-level' (4 or 5 ft. from the ground) is desirable, with a sunk ground floor.

Tower-mills and smock-mills require a stage around them just below the sail tips, to facilitate attention and repairs. These are found on the larger Lincolnshire tower-mills, and on Sussex and Kentish smock-mills, etc.

THE SPEED OF MILL MACHINERY

The layout of many windmills was poorly designed, partly because, in later years, they were built to a 'price' (sometimes under £500), owing to steam-mill competition. A good miller can usually suggest some obvious improvements to his particular windmill after a few years' experience. In the last fifty years of windmill construction (say 1840–90) it was not unknown for second-hand steam-mill gears and shafting, and pulleys, etc., to be built into new or reconstructed windmills, so that machines ran at unsuitable speeds, in an age when the approximately correct speeds had been pretty generally agreed by leading authorities from past experience. Such false economy could only hasten the end of the windmill as a thriving invention; but this did not apply to Lincolnshire, where many a good tower-mill was erected after 1840.

Sir William Fairbairn, who put engineering practice on a scientific footing through hundreds of systematic trials and experiments, laid down the following general figures for mill-work, more particularly for the water and steam corn-mills, of which he erected large numbers all over Europe and Asia.

Millstones, he said, should make 140 rev. per min. (he advocated 4 ft. stones—not 5 ft. 6 in. or 6 ft. stones as recommended by some of his contemporaries); dressing machine 350–650 rev., according to design and inclination of the sieve; bolting machine about 30 or 40 rev.; fan for millstones 560 rev.; elevators and creepers (conveyors) 50 rev.; sack-teagle, the same. Fairbairn was concerned with flour-mills; in a windmill or watermill for both flour and grist work the wheatstones were often geared to do 120/130 rev., and the barley stones 140. Sutcliffe (1816) put it the other way about—French stones 100 and Peak 75—but he favoured large 5 ft. stones for wheat and 6 ft. for grist. Hence the danger of ruining the meal at high speeds, for it emerges overheated (almost burnt) and overground from such big and cumbersome stones, unless they are driven very slowly.

(One old miller declared that it does not really matter which stones run faster; what matters, he said, is that the miller should dress his stones to get results *at the speed he expects the wind will drive them*, with the grain he proposes to grind. Another miller contended that if the stone-nuts are different sizes, thus giving different speeds, it is usually because they have been replaced, or because different-sized stones are, or have been, used in the mill concerned; but Mr Dallaway of Stone Cross Mill, Eastbourne, whose Peak stones have a small nut for faster running, said this is the local practice.)

An undated treatise on corn-milling, apparently published about 1865, gave 140 rev. for millstones; 60 rev. for the screws in wheat elevators; 130 strokes a minute for reciprocating dust separators and joggling screens; revolving screens and smutters for brushing wheat 470 rev. (with a fan at 550 rev.); worms for conveying clean wheat, and meal-worms and Archimedean screws 75 rev.; dressing machine 500 rev.; sack bollard 90 rev. and sack chain 160 ft. per minute.

Most corn-millers drive their sails at 12–15 rev. if the wind will do it; and marsh mills up to 20 rev.

It is asserted that an exhaust fan used in conjunction with millstones will enable double the work—6–10 bushels an hour—to be done, instead of 3–5 bushels; but doubling the actual speed of the stones calls for four times the power and is uneconomical, besides promoting excessive centrifugal force which might cause the stone to burst. A fan withdraws the dust and 'stives' from the grain into a 'stive box', and the finer particles of meal emerge from the stones more quickly, enabling the remainder to be better ground; whilst damp wheat is dried by the draught as it enters the stones and so will not clog them. The meal is actually discharged cooler with a fan than normally, if the stones are speeded up only within reasonable limits, and it can therefore be passed to the flour dresser after a shorter interval.

Marsh Windmills

Marsh-mills were built in the form of very large smock-mills 200 years ago, but were more recently constructed as small tower-mills or even open-frame mills without external casings. They usually contain only an upright shaft with a 'wallower' at the top and an inverted bevel pinion at the bottom. This engages with a 'pit-wheel' on the same shaft as the big paddle-wheel (15–20 ft.) which keeps the water moving in the dyke.

If the mill is required to lift the water instead of propelling it, a horizontal three-throw crankshaft is mounted midway up the mill, with long connecting rods driving an upright pumping engine. Square pistons were discovered by the writer in a mill of this type. These worked in a rectangular wooden chest, instead of having iron cylinders.

Chapter V

TOOLS AND EQUIPMENT

❧

'I N 1234 the King's Officer was ordered to let Henry Tyes take 30 logs in Grendon Wood to make a windmill....'

What kind of a windmill was to be fashioned from these logs? Perhaps they were only wanted for the main timbers, the smaller materials being already available, for 'logs' would undoubtedly refer to tree trunks or similar big material. No modern tools and lathes were at hand—probably only the woodman's axe and the carpenter's adze. With what skill and patience, then, did the men of old shape their timbers with the simplest tools—of this at least our medieval churches and cathedrals afford abundant evidence.

MEDIEVAL TRANSPORT

Removal of the logs from the woodland, if these early windmills were anything like as massive as their successors, must itself have presented a bit of a problem. Were they hauled all the way on the end of a rope, or did there exist timber trucks of some kind, with the solid wooden wheels that were still remembered in Lancashire in George Stephenson's time? Wheeled chariots were certainly used long before the windmill came to England; and it is known that wagons of some sort existed in 1198, a few years after the earliest recorded windmills. The pack-horse survived for centuries afterwards.

It seems likely that such timbers as could not be carried by horse were trailed at the end of a rope. This would enable difficult patches and ditches to be negotiated by slipping log rollers underneath the load, without the complications that arise when a loaded vehicle sinks into the mud. Even today one may see horses hauling tree trunks from the woods on the end of a chain, because the carts and motors are often unable to get over the quagmires that lie round gateways in bad weather.

MATERIALS DELIVERED

So we have a picture of these folk hauling logs out of the forest; per-haps first hewing them roughly to size in a clearing, and then making way to the windmill site through the muddy tracks of the cattle drover. Thereafter men laboriously struggled to the hilltop; urging on their teams with much shouting, and at last beginning to construct their windmill upon the highest point.

Before windmills ceased to be built the two-wheeled timber-bob had come into use—that handy contrivance which ploughed through the deepest woodland ruts with its great 7 ft. wheels and light frame, swinging its load on to road or track. There it was transferred to the four-wheeled timber truck or 'drug' by rolling the logs up two poles chained to fore and aft wheels, while a horse hauled at a double chain passed round the trunk and over the wagon.

What were the tools and equipment with which they might fashion and erect a windmill? Many of them are described in Henry C. Mercer's *Ancient Carpenters' Tools* (1929), and many of the following particulars are drawn from that source.

THE MILLWRIGHT'S TOOL-BAG

In the millwrighting craftsman's bag were a variety of special tools such as the compass plane with convex sole, for finishing the peri-phery of the great head and spur-wheels—a somewhat specialized millwright's implement, for neither Moxon (*Mechanik Exercises*, 1703) nor Holtzapffel (*Turning and Mechanical Manipulations*, 1843) describe it; the sun plane; semicircular stock and flat sole (a cooper's tool, for finishing rounded or barrel-shaped articles such as stone-cases); the twibill—a T-shaped tool resembling a small pick-axe (and called by the French *bec d'âne*—'donkey's nose') for hacking out mortices after drilling two holes with an auger—used from the time of Domesday Book to the seventeenth or eighteenth century, but now superseded by the manual-lever chisel.

Augers were always found there—the spiral auger, which Phineas Cooke claimed to have invented in 1770. The double spiral, for dis-

71

charging shavings, was incorporated in this tool, but it lacked the bottom router or screw (as used on a centre-bit) of the modern 'screw' auger. The latter, still in vogue for drilling bolt-holes through gate-posts and electricity poles and for the taking of borings by water boards, and so on, is unsuitable for a straight run of more then 2 or 3 ft. as the screw always turns aside with the grain when drilling a longitudinal shaft, no matter how truly the tool is lined up with the job.

Therefore the straight 'shell' auger with half-round section shank, and a little oblique cutting-edge on the tip, is sometimes employed for this work; but as it is very slow in operation, the millwright may prefer to compromise with the 'bull-nosed' auger, which resembles the 'screw' auger but has a box-like nose with two eyes or feed holes, that give it a curious owlish look (Pl. 20c).

All these tools had a T-handle, for turning with both hands. This gave them the advantage over brace and bit of much greater strength and leverage, and a reamer could complete to any desired exact diameter a bore thus begun.

For drilling small holes there was the bow-drill, with the bow-string looped round a bollard on the pointed drill, and the bow drawn back and forth like a violin bow; a variation of this was the thong drill, where one man held down the drill, round which the thong was coiled, and two other men drew the thong backwards and forwards.

MEDIEVAL LATHES

Obviously the ancient pole-lathe, dating back at least to the thirteenth century and surviving in our beech-woods till quite recently, was used for turning bollards, wooden spindles and so forth, and the wood to be worked upon was mounted between 'pikes' on a longish trestle, and a cord passed down to the treadle. Thus the stuff revolved to and fro while a tool was held against it.

Further improvement was obtained with the sixteenth- or seventeenth-century mandrel lathe, with its more complicated treadle and crank action, and belt drive.

Stone-nuts, wooden pulleys, etc., could be thus turned by wedging them on a pole or broomstick. For drilling through a pulley or spindle, the threaded feed which moved the mandrel towards the work was invented, it is thought, by Jacque Besson.

Finally, there was the great wheel lathe, described by Moxon in 1698—a 6 ft. cart-wheel mounted on a trestle and turned by hand-crank, with a strap round the tyre, driving a set of graded pulleys on the 12 ft. lathe bench. Alternatively, lathes were driven by water-wheels, and thus one axle-tree was probably applied to fashion another.

THE SAW AND THE ADZE

Millwrights had special long mortising chisels; and, for fashioning those great heavy timbers that are so important a feature of the long-lived windmill, they had the cross-cut saw, and rip saw, the saw-pit and the adze. The adze is rather an awkward tool for the beginner to use, though it is handled with surprising dexterity by the skilled artisan. It is shaped like a narrow, sharp-edged mattock, and is used for rapidly chopping away surplus material from the wind-shafts, main-posts and all big timbers, and for finishing them too.

In general, the saw does not seem to have enjoyed quite the hon-oured place that it enjoys today in the carpenter's tool-bag, except for the initial sawing of planks and posts in the saw-pit. But the invention of this tool is very ancient.

Timbers were often trimmed down to the proper thickness with an axe and draw-shave or adze, with a little planing if the face of the timber had to be 'true out o'wind', i.e. to present a dead-straight or square face. Projecting ends of tenons and wedges were chopped and chiselled off in preference to sawing; nevertheless, the old mill-wrights and sawyers knew well enough how to make their saws cut swift and true, otherwise the great long heavy baulks and planks they used would have entailed an inordinate amount of ultimate trimming and correcting to get them square along their length.

If they found a saw skewing off in a curve despite all their efforts to keep it to the straight, they quickly checked up on the teeth, which might be found to have a little more bias on one side than the other, and this they would carefully correct with a file, after a preliminary trueing up by punching each alternate tooth with hammer and centre-punch as the saw lay along a baulk of timber, or on the carpenter's stool or 'saw-horse'. This 'saw-horse', consisting of a squared log on four legs, they would certainly have with them on any important mill job; it was forever coming in handy for hammering a metal sheet, drilling a piece of wood, or a dozen other duties.

The straight lines which guided the saw, the millwright (like the carpenter) chalked by snapping his chalk-line upon the timber to be cut; and for checking a true face on a lengthy stock, a pair of 'trying-sticks'—square bars 6 or 8 in. long—were laid across the timber, one near each end, and a sight taken to see if they lay dead parallel to one another. Before the coming of spirit-levels, horizontal surfaces of timbers and millstones were checked with the primitive 'gable'—a rigid wooden frame, shaped like the letter A, with a plumb-bob suspended from its apex on a line which had to register with a central mark on the cross-bar of the A.

NAILS AND DOWELS

Although hand-forged nails were known to the Vikings, dowel-pins (sometimes called 'trunnels', or tree-nails, in their large form) found much favour for heavy timbers because of their permanence and rigidity under all conditions. Generally these were cut from the heart of oak—not the sappy portion—but Cobbett advocated growing the American locust-tree (the acacia) because, he said, locust pins would last ten times as long. Square pins used to be favoured (square pegs for use in round holes), but round pins are now more usual; and round pins were expeditiously prepared by driving the roughly trimmed sticks of oak through a special firmly mounted iron ring with a sharpened edge.

Small ironwork of many strange shapes went into the making of every windmill, especially post-mills; and the hand-made nails, for fashioning which the millwright might hire the village blacksmith's forge for himself for a day, were used by him to secure every conceivable kind of dog-iron, as well as straight iron plates or 'copses'.

74

It is interesting that these irons were cut from long bars with sledge and 'hard chisel' (even if half an inch in thickness) and not with the 'hack-saw' which millwrights today would consider indispensable.

Nor were nail-holes drilled in the irons, but punched with a smithy's sledge and square or round punch (the same as used for a hand-made horse-shoe), the iron to be pierced being taken red-hot from the furnace.

Auger holes in the timbers, if not clean and true, were finished by twisting a red-hot rod through them, followed, if necessary, by a dash of cold water.

The key-drift is a small tool for driving home the wedges that secure an iron 'cross' on a wind-shaft, or an iron spur-wheel on to the upright shaft, and similar jobs. It resembles a cold chisel, but the business end is bent, and has a blunt, slightly curved tip to provide a firm purchase while driving the key. Adjustable screw wrenches and spanners are often used on windmill repairs; but the author prefers an older type of wrench, 30 in. long, made from a bar of iron with a fixed jaw on the end; the sliding jaw is locked with a wedge that is slackened or driven home with a hammer; no screw adjustment or spring is required, nor does it loosen itself or slip off the nut as modern wrenches sometimes will.

HOISTING TACKLE

By such equipment mill timbers, fittings and accessories were all pre-pared; there remained the problem of assembling the fabricated materials into the integral framework of a windmill. Hoisting tackle used for windmill erection and repairs is brought to the site as soon as any major work is put in hand. Every sizeable part to be lifted is encircled with one or more sling ropes or sling chains, with small and large loops or links at the two extremities, the one being passed through the other. A similar sling is hitched around a trace-pole or the mill timbers, or whatever purchase point may be available; but sometimes if all the slings are in use, a short heavy rope may be substituted, and it will be tied in a bowline or 'bolan' knot. The millwright's stand-by, which he uses for every purpose and

75

occasion, this kind of knot (Fig. 6) cannot fail if properly made, yet the rope falls apart on being lightly pulled when disengagement is desired. To these two slings the two ends of a rope-tackle set will be coupled.

The tackle consists of a manilla or hempen rope, capable of holding 1 or 2 tons, and from 50 to 200 yards long, passed round double- or triple-sheave pulley-blocks, with one end secured to one of the blocks, and the other free for pulling. This multiplying gear gives a leverage of 4 or 5 to 1, and has an almost unlimited lift owing to its length. If the rope is properly unravelled by laying it down full length on the ground or by coiling it in a cylindrical pile before use, and if it is properly threaded on the pulley so as to avoid crossing, it is unlikely to jam or become entangled in itself. In weight it is light compared to its lifting power, but considerable time is spent in coiling and uncoiling it for each task. When using two rope sets simultaneously it is a good plan

Fig. 6. The bowline knot.

to have one larger than the other so that they can be called the big rope and the little rope, to avoid confusion which might lead to accidents.

CHAIN GEARS

For the heaviest timbers and shafts, main-post, crown-tree, windshaft, and so on, a stronger and lower-geared tackle is sometimes desirable; and the worm-and-chain gear blocking tackle as used in machine shops and garages is resorted to. But it is heavy and cumbersome to lift into place; it only lifts 10 or 12 ft. and therefore has not the required 'reach' for hoisting materials on to a mill; and the chains are easily entangled, causing much hindrance and inconvenience on outdoor work in windy weather. Such gears may have to be operated from midway up a ladder owing to the short control chain usually provided, rendering this, in view of the slight but ever-present risk of a ladder breaking, or of a chain snapping and letting the load fall against the ladder, the only hazardous part of a mill job in careful hands. The reduction ratio is very great, generally

30 or 40 to 1, enabling one man to lift slowly a load that four or five men could scarcely control with a rope-tackle.

For pulling materials horizontally, such as across the mill paddock, there is the endless chain pattern (no worm-gear or separate control chain) with a 12 or 15 to 1 differential gear formed by a double-block pulley with an extra notch on one side of its double-chain reel, so that the larger reel or pulley takes up or releases a little more chain than the other one (otherwise the chain would simply wind from one pulley to the other without moving the load). Technically known as the double-pulley chain blocking tackle, or the Weston differential pulley, it is shown in operation in Pl. 18. The chain feeds smoothly in a horizontal position with this gear, because it is the hauling and control chain combined, and is therefore in tension; but it is very dirty to handle if the pulleys are oiled, and it is not geared low enough to be used for heavy lifting jobs, nor does it incorporate a non-return ratchet.

Lifting chain-tackles are non-returnable by reason of the ingenious ratchet which enables the tackle to hold the load by itself. Normally a rope has to be held on to by all concerned, so long as it carries the weight, unless a post or tree is available round which to anchor it. Cranes are not favoured, since only a big one, reaching at least 50 ft. into the air, and comprising a costly item in itself, not to mention transportation costs, would meet requirements; and there is no sense in employing £1000 worth of equipment for a job which can be done with £100 worth. It is enough to say, perhaps, that in 1950 one millwright still conveyed his tackle (other than that essential adjunct to the rope or chain gear—the derrick or trace pole) in a 1928 Austin Seven, relying on any kind of ladder he could borrow at the site; and thus equipped he could tackle any windmill job in England.

Trace-poles or derricks, in the days when the construction of a new windmill was a familiar sight, consisted of a 60 or 80 ft. ship's mast, centrally placed at the new site, with its foot dropped into a hole in the ground and secured, maypole fashion, with strong guy ropes or wire hawsers to surrounding trees and gate-posts. But latterly, with only an occasional use for a derrick, a long mill-stock may suffice, as the tallest windmills have had their day, and are already dismantled or demolished. Even a pair of 30 ft. sail-clamps make a useful derrick for lifting a mainpost, etc. (see Pls. 18 and 19*a*).

JACKS

Another indispensable though rather cumbersome piece of equipment, which is brought to every major windmill job, is the lifting jack. Sometimes it is the old timber and iron affair (Pl. 20*b*), in which a crank-handle operates a rack and pinion, the rack having both a head and footstep, the latter being provided for jacking up a timber from within 6 in. of the ground level. A more modern one on the same principle, but of all-iron construction, is equally effective, but it is certainly an awkward and very heavy tool to drag about and set in position single-handed. More recently the Acrow-type iron pillar jack or prop, used by present-day contractors, has been introduced on windmill repair jobs. It consists of two long concentric tubes, the inner one being raised to its approximate setting and secured by a peg in one of a series of holes, whilst the final lift is applied by a lever-operated collar or nut which screws up the head of the jack. This is particularly suitable for shoring up a mill or its floors; its actual *lifting* power is limited by the short operating lever and coarse thread.

Alternatively, where moderate lifting power is required for trueing up sagging floors, or levelling a wind-shaft and similar jobs, the hydraulic lorry-jack comes in handy, particularly in confined spaces, and it is lighter to handle and more convenient to stow away on a vehicle.

REPAIR CRADLES

Although a few well-equipped windmills on which the miller undertakes some of his own repairs have their own cradle for repainting the mill walls, the millwright generally keeps his own cradle to be taken to the scene of work as required. It will usually be 10 or 15 ft. long with a double rail all round (but some millwrights prefer a single low rail so that they can step over the side easily to slide down a rope to the ground or to climb in and out of the mill window), and there are wooden wheels about 9 in. in diameter at each end of the cradle, placed to run up and down the weather-boarding so that the framework of the cradle does not catch under the edges of the boards when ascending. Two sets of rope-tackle can be used for suspending the cradle from a cap stage; but some millwrights favour plain ropes on a post-mill, simply securing them over the rope shoes (if provided) on the roof, and into the far window (where they are tied to the main-post) or down to the wheel of a cart placed sideways on, in the yard. Small one-man cradles, as used at Upminster, are kept mainly by millers for minor repair work. All cradles should have a separate safety rope to be hitched on each time the cradle is raised or lowered.

Chapter VI

BUILDING A WINDMILL

❧

THE men who designed, fabricated, erected and opened the Crystal Palace in eleven months from the day when it was roughly sketched in on a sheet of blotting paper must have been good organizers; and the same may be claimed for those great engineers who brought a network of railways to every corner of Britain within the short space of fifty years, and for the engineers who in seven years erected the world-famous Forth Bridge. Capable organization was the rule, not the exception, in the eighteenth and nineteenth centuries, and evidence of careful organization at a windmill site may be found in the short times known to have been occupied in erecting various post-mills up and down the country—for example, Banner Mill, Quainton (Buckinghamshire), which was started on 16 May 1797: roof boards on, 16 June; sails up, 26 June; and the first sack of flour ground on 20 July—nine weeks in all.

The weightiest timbers for the new windmill would have been deposited conveniently around the mill yard, or in the mill field, so that they could readily be brought within reach of the lifting tackle by rolling them on logs, or turning them over with pinch-bars, or if necessary pulling them with a heavy rope coupled to a horse or to a block-tackle. Everything was prepared in advance so far as main timbers and mill-framing were concerned, such items as the sides of the mill-body probably being temporarily assembled and dismantled; but the floor-boards and weather-boards would be cut to length on the job, and the small fittings installed.

SHAPING THE MAIN TIMBERS

No man now living has made the main timbers of a post-mill, though the late Mr Amos Clarke, who died in 1953, saw his father cut down an oak tree and fashion from it the last main-post ever made (and also the crown-tree from another oak nearby) for Wetheringsett Mill near Stowmarket in 1883.

In shaping the main-post, it was first necessary, after cutting it to length, to decide upon the centre-line of the post and to insert gudgeon pins at each end, which would rest in V-grooves on the trestles set up firmly for the job. The post was slowly revolved with a large crank-handle while the cutting tool was applied to the work.

The big recesses made to straddle the cross-trees at the foot of the post were fashioned by sawing up the sides of each recess from the bottom, then drilling several lateral holes through. The material between these bores was chopped out with long mortising chisels until a slot was opened through to accommodate a large saw, with which the cut was finished and the slab of wood released.

It was then necessary to cut the quarter-bar mortises; and there is little doubt that the exact angle of the quarter-bars was predetermined on paper, and the tenons and mortises at the confluence with cross-trees and main-post were fitted with the help of templates to ensure the accurate placing of each quarter-bar. Any slackness after the post and quarter-bars had been assembled would mean scrapping and replacing at least one of the bars.

Thence to the crown-tree: the socket in it would be tested against the pintle of the main-post, the side-girts compared to the rebate in the crown-tree, the corner-posts temporarily fitted to the side-girts, and so on, thus making sure that all joints would fit snugly when finally assembled.

In erecting the very earliest windmills, the corner-posts or main-post (whichever they depended upon) would be planted securely in the ground; for it was probably only later that millwrights learned to mount a post-mill free of all retaining irons upon its piers. A fixed mill they would build much as we would erect a barn today; but a revolving post-mill, after the main-post had been set, would be constructed about its crown-tree, to the ends of which the side-girts, if it had any, were secured.

ERECTING A POST-MILL

Generally speaking the method of erecting a post-mill as we know it today, with brick piers above ground, would be as follows:

Supposing the piers to be several feet in height, one of the cross-trees could be hoisted first one end and then the other by several men standing on the piers with ropes; but for very high piers, such as are sometimes used in East Anglia, a pole might be lashed to each pier, from which to suspend a rope-tackle for this purpose. A central derrick-pole (such as an old ship's mast) could be erected by dropping one end into a hole and raising the shaft or pole. Guy-ropes for the derrick would then be secured to convenient gate-posts or trees round the mill yard.

THE TRACE-POLE

The foot of the derrick or trace-pole can be either lashed to a sail-stock or dropped into a hole to prevent kicking when taking up the load. When the first cross-tree is lifted, it can be settled on the lower pair of piers, and the second cross-tree swung round over it; or the first could be put on the higher piers, and the second slung beneath it, as there ought to be an inch clearance at the intersection. And if these cross-trees are set in place first, without the aid of a derrick pole, they will assist the placing of the derrick pole for subsequent use, by ensuring that the pole is as central as possible yet not in the way of the cross-trees.

It is a day's work for two or three men to erect a substantial trace-pole, reaching 15 or 20 ft. above the piers, and rigidly secure it with guy-ropes, getting the sling chains and lifting tackle properly placed, and manoeuvring the main-post into a convenient position for lifting in such a way that it does not foul the cross-trees and yet can be induced to set down in the right position when suspended above them. It must be remembered that the cross-trees and post must each face a prearranged way round in relation to each other, as the quarter-bars, being individually fitted, are not likely to be interchangeable.

HOISTING THE MAIN-POST

When all is ready, sling ropes or chains are fixed round the waist of the post, which will right itself when lifted, as the bottom half is heavier than the top (Pl. 19 a). But the strongest available rope-tackle will be wanted to lift the head of the post first if the chain-gear suspended from the trace-pole will not reach the ground. At the earliest opportunity the chain-tackle is coupled up, and a 2-ton gear is desirable, although up to a century or two ago, only rope-tackle and wooden block-pulleys were available.

Great caution and judgement need to be exercised in placing a ladder for operating the chain-tackle, as the main-post may swing about when clear of the ground; for this reason in fact the millwright might prefer to employ the extra labour entailed in lifting the post solely with a rope-tackle.

Next, two of the quarter-bars must be hauled up on to the first (upper) cross-tree, and their feet set in the mortises provided for them.

Then comes the ticklish job of inducing the heads of the two quarter-bars to engage with their respective mortises in the main-post as it is very gently lowered. This may call for a good deal of manoeuvring with crow-bars and ropes; at this juncture the post is difficult to get at if the cross-trees are 10 or 12 ft. from the ground; and again the coaxing may have to be done from ladders placed within range of the suspended main-post.

However, tens of thousands of post-mills have been erected in the last 500 years, with more primitive equipment than is now available; to a capable millwright such problems were all in the day's work, and a score of men in any village could be found to help him to erect the mill, when skilled labour had cut and shaped the timbers.

THE CROWN-TREE

Having successfully mounted the main-post, it may be necessary to rearrange the lifting tackle as the sling chains now need to be 2 or 3 yards above the top of the main-post so that the 30 in. thick crown-tree can be slung well above it, for lowering on to the pintle. An average crown-tree can be hoisted by six or eight men with the heaviest rope-tackle, the chain-gear being hardly essential, since this timber does not require to be suspended for any length of time if the job goes smoothly.

Both ends of the crown-tree may now be lashed down to the cross-trees (or propped from the ground) to maintain equilibrium as successive timbers are installed first on one side then on the other; in fact, the windmill can be made to form its own derrick-pole as the work proceeds.

FRAMEWORK OF THE BODY, AND THE STEPS

It is well to have the sheer-trees (beneath the body) erected at this stage; they are about as heavy as the cross-trees, being a bit stouter but shorter, and can be lifted with two sets of tackle slung about the centre of the crown-tree; they should be supported by tie-rods reaching down from the crown-tree on either side of the main-post. If a square collar or 'girdle' is provided, this can be fitted while holding the sheer-trees a little above their final position. The collar would be in the way when hoisting the sheer-trees but, if not provided, it may be advisable to level the sheer-trees temporarily (by inserting wedges over the shoulders of the quarter-bars) and brace them together.

The bottom rails of the head and tail walls may come next, resting upon the ends of the sheer-trees; then the bottom side-beams, and the flooring, followed by the side-girts, which rest on the crown-tree.

If the central trace-pole is not very tall, it may be an advantage to erect two poles, towards the ends of the crown-tree, to secure a more direct lift for each side-girt (the ends of which will be coupled to rope-tackle sets, the same as for sheer-trees). These erected poles would help to support the crown-tree on an even keel; but the side sheers, resting on the inwardly inclined rebates of the crown-tree, will hold level enough when tie-rods are inserted through sheers and tree, from above, down to the bottom side-rails of the floor.

The corner posts, the meal-beam connecting the forward ends of the side-girts, and the central prick-post which will help to support the weather-beam and sails, can then be established in their respective places (see Tunstall Mill, Pl. 13 *d*).

The side-girts do not need to be fixed to the crown-tree with dog-irons (as practised in the south Midlands) if they are properly set on the rebates, with upright tie-rods; but the weather-beam must always have strong and well-placed irons. The head-room on the lower floor is greater, especially in the north Midlands, than on the upper or meal floor which lies between the side-girts, therefore the corner posts can be gripped at the waist with the 'gear' and induced to stand upright. Trace-poles at either end of the crown-tree would lift the top-rails at the eaves if tall enough; and the roof-studs and ridge-beam could be assembled by hand.

All this framework needs to be temporarily strung together, because there are the intermediate studs and diagonal braces to be mortised-in (as provided for) before the main sheers and rails can be drawn together for final dowel-pinning and bolting.

Sometimes, abroad, the lower half of each side-wall has been finally assembled on the ground and hoisted with some kind of derrick crane, followed by the complete head and tail walls *simultaneously*, and then the upper side-walls; but English millwrights did not reckon the use of such heavy and expensive equipment economical, and (at least so far as the writer knows) always assembled their post-mills piece by piece.

Now the very cumbersome weather-beam can be hoisted, preferably with a derrick-pole lashed to the inside of the prick-post, as a central trace-pole would need to be very tall and strong to lift this weight up the front of the mill.

Floor-joists and stone-beams or bearers will be inserted while the heavier jobs are progressing, and the door-posts and lintel and the window-frames and roof can also be fitted—straightforward builder's work for the most part. The stair strings and treads can then be assembled on the ground and lifted with a rope-tackle slung from the door lintel, which should be lashed back to the main-post to take the strain if the ladder be long and heavy (Pl. 20*b*).

Most of the detail construction can now proceed—the steps can be raised at the foot with a 5-ton jack whilst the tram-frame and its wheels are attached; the tail-beam and the upper side-girts which support it can be installed and the roof boarded. It will be useful to have steps attached now to turn the mill, and to improve the balance while the sails are being fitted; also the steps provide better access to the interior.

THE WIND-SHAFT, BRAKE-WHEEL AND BRAKE

Now we are ready for the wind-shaft; and the millwright will have his own method of hoisting this. The commonest method is said to have been to haul the shaft up the breast of the mill and then swing it in tail first. For this purpose a trace-pole formed of a very long sail-stock can be roped to the weather-beam (which in turn is lashed to the crown-tree), the foot of the pole being secured to prevent kicking. Alternatively, a ship's mast 70 or 80 ft. high may have been put up in front of the proposed mill, where it would be of general service, besides assisting in this final operation—this was a favourite practice with some millwrights, even for hoisting a 'pre-fab' top on to a tower-mill.

A wind-shaft, being very heavy, will be hoisted with two sets of ropes, getting its tail end over the sill first and hauling it in with a Weston-type chain-gear, whilst a heavy rope coiled round the main-post within the mill holds the tail of the wind-shaft in a downward direction to prevent it smashing up through the roof (Pl. 20a).

An alternative method is to use two poles 50 or 60 ft. high, meeting at the apex like a giant pair of sheer-legs, setting them up at a distance of about 10 ft. in front of the mill and retaining them in position with guy-ropes. Now hoist the wind-shaft up to the requisite height and then allow the sheer-legs to recline slowly back towards the mill (meanwhile coaxing the tail of the wind-shaft over the weather-beam). This is done by gradually easing off the guy-ropes ahead of the sheer-legs and taking in those to the rear, so that the whole contrivance eventually comes to rest against the front wall of the mill. These two poles or sheer-legs could be assembled at the outset either by lashing them together whilst resting against the mill, or by coupling them on the ground and then hauling them up.

If the head-wheel or tail-wheel is in one piece, or if it is otherwise thought desirable to assemble them on the ground before installing, they will need to be suspended from the roof (specially strutted for the purpose) while the wind-shaft is passed through them; when these are approximately in place, the nose of the shaft, with the aid of the trace-pole (or a small pair of 'sheer legs' mounted on the weather-beam), will need lifting while the 'neck-wear' is inserted and bolted down beneath it, as this would be in the way when pulling the shaft over the sill. 'Compass-arm' brake-wheels (see Chapter 3) are necessarily assembled *after* the wind-shaft is installed, as the arms pass right through very deep mortises in the shaft, and are half-housed together; the remaining apertures in the mortises are subsequently blocked with several wedges.

It will be convenient to fix the brake next, getting it fully adjusted and in working order if possible.

HOISTING THE SAIL-STOCKS

Now the sail-stocks can be brought up and erected. In a mill yard, if half a dozen strong men and a heavy rope-tackle are available, a stock may be hoisted without a chain-gear. With a long heavy stock on a tall mill, however, it would be better to engage more men on the rope and to lash a pole on to the sail-stock to facilitate guiding it from the ground as it is directed into the canister. At this point much aggravating delay may be occasioned by the stock binding in the box, and ropes attached to its ends may have to be plied vigorously and repeatedly, to coax the timber towards its final place.

When each stock is nearly home up to the shoulder, it is pulled round top to bottom, with a rope on the tip, so that it slips home and can be wedged in the poll-end. The stock is easily lined up for wedging, by setting it horizontally, looking along it with a good eye, and signalling the helpers with ropes to pull back or forth. The wedges should be in front of the outside stock and behind the inside stock, so that the sails 'lay together', as the millers say; and with the sail stocks in place, the 'striking rod' can be pushed home ready for coupling up.

FIXING THE SAILS

The first sail is laid with its heel pointing to the mill and is lifted with rope-gear only (using a sling rope—not a chain—round the sail-bars, as this does not bite the wood so much). Two men hold back the tip of the sail, using ropes if it is a tall mill, to make the heel ascend in front of the stock instead of getting behind it. Little difficulty arises in manoeuvring the sail to its appointed place; and the bolts and straps are fixed by climbing the sail; this is safer than using a ladder, in case the sail slips a little and upsets it.

If the sails have been matched up with the stocks on the ground, in order to fix the straps exactly, rebating the wood where necessary, all four sails could be installed by four or five men in a day; but if it is found impracticable to lay out the sails on the stocks in a restricted yard, the bolt-holes can be drilled by measurement, and the straps fitted during erection. Back stays from sail to stock should be fixed as soon as possible, in case a gale comes up.

Sail shutters for 'spring' or 'patent' sails, are hauled up three at a time, by a man on the sails (the thimbles having already been screwed on, up one side of the frame), and duly fixed and coupled by their 'levers' to the sail-rod which operates them. The 'triangles' can now be mounted on the stocks and coupled to the sail-rods and also to the 'striking rod', which is partly passed through the wind-shaft before hoisting the latter on to the mill.[1]

THE FLY-TACKLE

The fantail or fly-gear may then be erected—the big fly-posts are bolted to the foot of the steps, with their cross-braces lying across the stairway, and lifted into an upright position by means of the big jack. A rope can be hitched from mill-body to fly-posts to prevent them swinging too far back when raised; and after fixing the fly-

---

[1] When converting from 'common' to 'patent' sails, wooden wind-shafts have been bored by lining up a 'bull-nosed' auger (Pl. 20c) on jigs on the fanstage and setting the sails off to turn the shaft against the auger. An iron shaft has been similarly drilled from end to end with a 'D' bit.

strings to secure the posts in position, the complete fly on its iron spindle is hoisted by a rope-tackle attached to the arm that carries the pinion, as this end of the spindle is heavier than the other. It then remains to connect the driving rods which couple the fan-pinion to the tram-wheels, but this might be left for a bit, and the fly chained up until the sails are ready for use.

COGGING THE MILL

But while all these many and various preparations are going forward, an important self-contained job is being done by an expert amongst experts—a master craftsman of the countryside is making and fitting the many wooden cogs in the great wheels as they come to him from foundry or joiner's bench. For these a stock of 2 in. apple- or beech-boards was kept in hand by old-time millwrights; but such watermill cogs as were continually exposed to moisture were of oak.

The cutting of anything from 70 to 140 mortises in a big wooden wheel is one of the most skilled arts of the millwright, always presenting pitfalls for a careless workman. All need to be precisely the same size and shape, and all evenly spaced to a hair's breadth; for a faulty mortise will not hold a cog for long.

When cogging a wheel in the workshop, it is convenient to mount it keyed on its axle with one end of the axle resting in a block secured to the doorpost, where plenty of light falls, and the other end on a portable bench, so that the wheel may be revolved in a true circle as and when desired.

An iron wheel, or an old wooden one, will be mortised already, but a new wooden one will next require marking out for this purpose. A rigid horse is set alongside the wheel, to take a scribe or marker, and an improvised crank-handle is tightly fitted on to the axle, to turn the wheel so that the marker will describe a circumferential line along which the cogs are to be spaced. The exact centre-line of every mortise is then marked off very carefully upon this line by setting a sharply pointed pair of dividers to the dimension previously calculated to give the required spacing, and working carefully round the circle till the starting point is reached again. If there should be a small error at the finish, the dividers will be adjusted accordingly, so that eventually, and not without a good deal of trouble, the correct markings are established.

Now the full outline of each mortise can be set out; and the corresponding outlines on the back of the rim (where the shanks of the cogs will project) are traced round it with a try-square. The mortises are drilled with an auger, then chiselled with the broadest practicable chisels, bearing in mind the taper of the mortise; and little discrepancy will be found if a good tradesman has been deputed to the task.

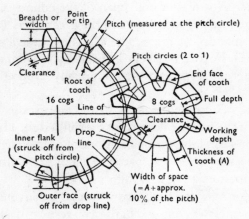

Fig. 7.   Detail of cog-wheels.

Fitting 'Blank' Cogs

Cog 'blanks' are sawn from a heavy plank, preferably apple-wood, which must have been felled over nine years (or the cogs will eventually shrink), and the shanks of the cogs are accurately sawn to shape. All the cogs are driven into their mortises up to the shoulders with a heavy mallet; and any cogs found to be slack or easy-fitting are scrapped and done again. A trimming tool is now set in the stand, to turn up the two end-faces of all the cogs and give them their true breadth (see Fig. 7).

Finding the Centre-lines

The required *depth* of the cogs is next marked on their ends with a scribing tool while the wheel is turned; and they will then be sawn off accordingly, leaving only the contact surfaces or flanks of the cogs to be fashioned by describing the requisite outline or profile on the end of each cog in readiness for chiselling off. For this purpose the 'pitch-line' and the 'drop-line' are engraved on the ends of all the cogs (Fig. 7) by a 'gouge' fixed in the stand. The *pitch-line* is explained in most mechanics' handbooks; and in the case of a spur-wheel and pinion, the pitch-lines or circles in the two wheels to be meshed are proportional in diameter to the respective number of cogs, and they touch at the 'line of centres' as shown in Fig. 7.

Shaping the Contact Surfaces

The verges of the tips or faces of the cogs, on a spur-wheel, can be inscribed with a try-square, since the butt of the tool will rest squarely on the ends of the cogs; but on a bevel wheel, these are on a curved plane or frustum, and the margins of several cogs must be established by striking off opposing arcs from the corresponding corners of two or three *alternate* cogs, so that they intersect near the inner edge of the frustrum. This will locate the outline of the *intermediate* cog, and so on, but to save repeating all round the wheel, the adjustable scribing arc (Pl. 22 *a*) is set so that its end-plates rest upon the shoulders of two cogs, whilst its central arm is lined up with the marked edge of the intermediate cog, thus enabling this marking to be constantly reproduced.

Now the surplus wood can be sawn away, leaving the flanks to be trimmed down to an exact curvature with a broad gouging chisel for the inner flank if this is concave, and a plain one for the outer face. These chisels should be extremely keen-edged because the wood is hard, and they must run true and parallel with the outlines across the breadth of the cog, or the work will be ruined; and the skilled craftsman can fashion a practically perfect set of bevel cogs without plotting the profiles of their small or inner ends, trusting to the eye to cut an evenly tapered surface from end to end.

91

So tightly should all the cogs hold in the mortises, without fixing, that all this work should have caused them no disturbance; but they must now be pinned through the tail-ends of their shanks, by drilling every cog through its breadth to receive pins (which might take the form of 4 in. or 5 in. nails) as required. The greater the breadth, incidentally, and the more numerous the cogs, the higher the power they will transmit without damage, because of the broad bearing surface; but thick short cogs, which give rise to more noise and vibration, are easier to make.

The re-cogging of the 4 ft. wheel illustrated (Pl. 22a), with about 80 cogs, was expected to occupy 3 weeks, including dismantling and removing wheel and shaft to workshop and reassembling.

Having finally installed the wheel in its appointed place, it only remains to set the mill going and *listen*—any faulty cogs will soon be heard. Mr Walter Rose of Haddenham mentions in his book *The Village Carpenter* (1937) that his people, true to old-time tradition, dispensed with the nails in favour of the alternative method of driving a wedge between each pair of shanks (Pl. 22b) and he was proud to say they never heard faulty cogs, nor did their cogs come adrift.

Inaccuracy increases with use; and if one cog breaks, several more may go each time round, before the mill can be stopped.

INSTALLING AND BALANCING THE MILLSTONES

The millstones can be hauled up the steps, lying flat, and turned on edge on the threshold to get them in—a ticklish job, as a slip might send the stone overboard and perhaps smash a quarter-bar or kill someone. Probably the stones will be handled with gear suspended from the crown-tree, and they will be parked on the lower floor until the wind-shaft, offering sufficient purchase to lift them to the upper floor, is installed; then they can be lifted through the special aperture provided by having a removable section of flooring. When the sails are on, a rope wound round the wind-shaft will lift the stones.

A new bed-stone is once said to have been collected from the railway station, installed in a mill, the runner-stone and mill gear replaced, and some grinding done, between 9 a.m. and 5 p.m. (see Pl. 28 a).

New runner-stones are more trouble, because they have to be balanced by trial and error on their spindles and maces, the cross-bridge-piece perhaps requiring resetting with lead melted in a 'stun-set'.

PAINTING A POST-MILL

Painting a white-walled post-mill is accomplished first by sitting on the top ridge, then by entering a cradle from the sail stocks, later entering it from the upper window, then from the lower one, till the bottom of the body is reached. The cradle is secured by two ropes passed over the iron 'rope-shoes' which should be provided on the roof, and tied to the wheel of a cart on the far side of the mill; although they are sometimes passed into the farther window and secured to the interior timbers.

About 3 cwt. of paint (two coats) is required after priming, etc., for a white post-mill; and painting one side takes two men a day, using a full-length cradle.

The sails are painted by two men climbing up them to do the face; another two paint the back from a cradle or ladders, or even from the backstays; and to cover a full set of double-shuttered 'patents' with one coat is a good day's work for the four men. Those working on the front face should begin first (at the top), as some of their paint will probably drip through, owing to the slant of the sails; it is therefore advisable for the men behind to be a little higher up all the time.

AUXILIARY MACHINES AND THE ROUND-HOUSE

Meanwhile such items as the flour-dresser, sack-bollard, and feeding spouts and hoppers can be built in; the weather-boarding will be finished off, and the bricklayers will construct the post-mill round-house, which would only obstruct operations if built at the outset.

Someone will be trueing up the big wheels by adjusting the wedges and inserting slivers between the arms and felloes, trimming faulty cogs, building in the grain-bins, fixing the windows, regulating the sail-shutters, fitting the reefing gear for 'patents', laying the tramway of stone, timber or concrete (or even iron plates occasionally), fitting belts, adjusting tenter-gear and governors, and packing bearings with 'black grease'. At last the mill is in perfect trim and all is ready for filling the grain-bins and grinding the first sack of corn—a red-letter day in the history of any windmill.

I have gone into some detail about the erection of a post-mill. Now I shall not confuse the picture by leading on to smock-mills and tower-mills in detail; but I should like to point to an important discovery not yet mentioned in print. The late Mr Hector Stone of Upminster said that instead of erecting the eight corner-posts of a smock-mill one by one in the usual manner, they sometimes constructed the four alternate sides of the mill on the ground, and laid them with their heads facing the mill base. Each of these sides was then hauled up into its place and secured with guy ropes, etc., and the four intervening gaps were filled in with suitable framing. This is in my opinion confirmed by the discovery that four alternate sides of Lacey Green smock-mill are about 6 in. wider than the other four—a peculiarity that is otherwise difficult to account for.

Unlike post-mills and smock-mills, tower-mills were built with scaffolding, but in some cases this was used *internally*, as in erecting factory chimneys, and not externally.

Chapter VII

THE WINDMILL AT WORK

❧

A LOT of people in this country today have never seen a windmill at work; and, indeed, many have never been inside a windmill at all, working or idle. Now and then it has been my job to take people over a windmill on a first visit; they never fail to exclaim with surprise at the quantity of fine timbers and the superb workmanship—especially in, say, things like wooden cogged wheels. The intricacies of an elaborately fitted mill-cap with its centering and running wheels, its fantail mechanism, and its huge brake-wheel and brake, are greeted with exclamations of astonishment—astonishment that such a maze of detail could be so ingeniously fitted into so confined a space. The visitor's curiosity is certainly aroused; but as often as not it is impossible to put the windmill to work, even if an experienced miller is still available.

STARTING THE WINDMILL

One is often asked how a windmill is set to work; and the preceding chapters have to some degree answered the question.

When the miller goes aloft to begin his day's work, the stones may have been 'laid to'—or they may be disengaged so that the sails can idle round in the breeze to relieve them of strain—they will not normally run too fast with cloths furled or shutters open. A glance will tell whether the bins and hoppers are charged with grain; tentering gear (see Chapter 3) and feeder-shoe may need adjustment if a new batch of meal is to be ground, and the 'spattles' on feed-hopper and meal-spout may need opening.

If the mill has been left free, the brake may now be applied and the appropriate stone-nut engaged, and perhaps the dresser or some other machine put into gear. The sails must be braked in any case if of cloth, because the next item will be to reef them, after first levering up the steps and pushing the mill round into the wind, for the heavy cloths are unmanageable with the wind behind them and may be torn off the sails.

The long brake-rope is thrown out over the side of the mill or let down through the floor, and the miller descends to the yard and spreads one of the cloths with the aid of the sail-ropes (Chapter 3). Then he releases the brake sufficiently for the next sail to come down, and stops the mill again. Each sail is thus spread in turn, and the brake is finally released. (Should the wind, playing on the sail-frames, be insufficient to bring them round for reefing or unfurling, it may be necessary to pull them round with a boat-hook or pole, but this is rather difficult if there is only one pair of sails and they are resting horizontally.)

SAIL ADJUSTMENTS

'Spring' sails must be brought round one at a time, as 'common' or cloth sails are, for setting the tension lever of each spring; sometimes a travelling platform like a wheeled table has to be trundled round the mill for this; but 'patents' are started from outside the mill by adding weights on the chain to bring the shutters up to the wind, and releasing the brake; and being easily controlled they are a great asset. But some old millers preferred the fast running of cloth sails—about 8 revolutions or 'bouts' of the stones to one of the sails, against 11 or 12 revolutions of the stones with patent sails—because they liked to see a mill running with sails swinging round at a good pace.

Because the work never had to be interrupted for reefing, patent sails easily did more work on a given mill than cloths, in spite of being heavier and presenting a less solid surface to the wind so that they did not 'strike' so readily in a light wind. With tower-mills, cloth sails were rather uncontrollable unless a very large cap was provided to accommodate a big brake-wheel for stopping the mill.

In action, spring sails differ from patent sails in that the shutters of individual sails adjust themselves during each revolution, opening a bit at the top of the circle and closing towards the bottom with a snap which sprays water over the wall of the mill in wet weather (hence the sheet-iron covering on the breasts and sides of so many Sussex post-mills). The shutters of all the patent sails of a mill regulate simultaneously; but a badly proportioned striking gear will cause the weights to swing violently up and down with the gusts of wind, instead of rising and falling smoothly.

FILLING THE GRAIN BINS

With sails all set, the next task, if the bins need replenishment, will be to bring the newly delivered sacks up from the round-house or ground floor, dropping the chain through the openings in the sack traps, if it has been coiled overnight in case of lightning, and hooking it round a sack. Returning aloft, the miller pulls on the control rope to engage the bollard, retaining a gentle hold with one hand and steadying the sack with the other as it breasts the edge of the bin, where he balances it whilst disconnecting the chain, thus obviating 'man-handling' when emptying the sack. By uncoiling a few feet of chain the chain's own weight will bring the chain back till it once again comes to rest on the store of sacks in the round-house; in the absence of a mill-boy the miller again nips down the stairs to hitch another sack on.

A more fortunate miller, who may have at any rate an hour's help, morning or evening, from his son or some other lad, will teach him to employ a 'morse code' on the control rope—one jerk, 'pull up'; two jerks, 'stop'; several, 'all up, no more sacks to despatch'. With a tolerant miller a small enough lad might even help himself to a last ascent on the end of the chain.

If the sack-hoist drive is by belt, or even a chain on a plain grooved drum, the miller can conveniently lower his freshly loaded sacks of meal to the ground by slipping the belt, maintaining just sufficient tension on the hand rope to cause the belt-drive to check the release of the chain from the bollard, so that the sack descends at a controlled rate. But a gear-driven bollard does not provide for lowering the sacks. This is a considerable handicap.

Whilst thus busily working, the miller gives an occasional glance at the clouds from the window or 'clap'; unless he has a fantail, when the mill is automatically kept in the eye of the wind, and the miller need not worry about changes of wind.

TENTERING GEAR

In getting the mill in good trim when starting on the day's work, much will depend on the layout of tentering gear and steelyards, and the miller's skill in adjusting them; for if badly designed, the governors will not regulate the stones satisfactorily in choppy winds. Separate governors for each pair of stones are desirable except in big mills where long and well-placed steelyards can be provided, and these should have closely spaced notches, say half-an-inch apart, not several inches, or fine adjustment is greatly hindered (see Fig. 5).

After having had the stones up for redressing, a day or so will probably be spent in getting the tentering gear nicely adjusted again; but grist-milling calls for little further adjustment. On more important work, however, it may take the first hour or two of the day to set the stones to the best advantage for producing a good texture of meal, particularly on an old post-mill in which the stones are inclined to settle and bind when she shifts into a different wind. Now and again one hears of an old tail-pole mill on which the stones needed re-trimming every time she was pushed round.

Particularly aggravating are windmills in which the 'gimbals' supporting the runner-stone rise on the trunnions of the mace owing to excessive wear (see Chapter 3, 'The Mill-rhynd') as it gathers speed—a thing it cannot do with a 'bridge'—and it is probable that some millers never discovered the real cause of this, although put to endless trouble in screwing down the stone and letting it up again as the wind rose and fell. Some 'gimbals' have properly recessed lugs which will not rise on the trunnions of the mace and are consequently free from this particular annoyance.

HURST-FRAME AND SPUR-GEAR POST-MILLS

The change-over from 'head-and-tail' to 'spur-gear' (see Chapter 1, p. 12, 'Layout of Millstones') which provided one-and-a-half clear floors, was much appreciated, especially in smallish mills with lack of head room, where the big exposed wheels and the millstones had filled the whole working floor so that manoeuvring about to make adjustments was irksome and dangerous and even led to scalped heads and severed arms if the millers were unwary. Conversion to a 'hurst-frame' layout, however, with gears and millstones all concentrated on the lower floor, was a mixed blessing, for though it cleared the top floor it prevented sacks being properly suspended beneath the millstones, so that the miller had to be ever in attendance to shake down the meal—and a bin in which several sacks of meal could be accumulated and mixed for even texture was practically out of the question.[1]

TYPES OF MILLSTONE

Millstones have to be re-dressed (nowadays by the miller) at fairly frequent intervals, depending on the quality of stone and the dressing employed. Barley-stones (which grind perhaps one sack per hour in a light wind, or three or four with a strong wind), are of rough Derbyshire Peak stone in one piece and these may need re-dressing every fortnight. Wheat requires a hard smooth-faced stone, and a French burr-stone is used for wheat; it is also suitable for grinding oats, beans and maize. This stone does not occur naturally in very large pieces, and a burr-stone is therefore composed of several prepared blocks cemented together (bound with iron hoops which are riveted at the joint and then shrunk on) and mounted in a base of plaster of Paris, sometimes mixed with Portland cement. A burr-stone is harder to dress than a Peak stone, but it should only need freshening-up about once a month. Some burr-stones are built into previously prepared hoops.

---

[1] Bozeat Mill was reconstructed with the millstones ahead of the brake-wheel, which obviated the conversion from left-hand to right-hand sails, as well as saving space.

LIFTING A MILLSTONE

To re-dress the millstones the runner-stone must first be upturned; it is raised (unless there is some lifting tackle) with a strong pinch-bar, and a big triangular wooden block is inserted beneath. Two stout wooden rollers about 18 in. long are slipped under, one on each side, in line with each other, supporting the stone high enough to clear the mace, so that it can be slid horizontally off the bed-stone (with plenty of bran between, in case it 'flops' and hits the lower stone) on to some sacks of straw, or a pile of empty sacks.

The triangular block is again used if necessary to support the stone while slinging a very substantial rope round it, through the eye; and this rope may be hitched to the wind-shaft to make the sails lift up the stone on edge. Then it is slewed round and very gently and cautiously lowered back on to the bed-stone (or on to the floor if there is room for it), face upwards, ready for dressing, re-membering all the time that a slip might bring a ton of stone down on somebody's foot.

A few windmills are furnished with a 'stone-crane', consisting of an upright iron pillar with a large swinging bracket, from the end of which bracket are suspended a pair of curved arms each terminating in a hook or lip. These two hooks are inserted in sockets already provided in the opposite sides of the stone, so that when lifted the stone is symmetrically balanced and can therefore be swung over, face upwards or downwards as desired, and the hinged bracket will put it down where it is wanted. The same can be done with a worm-and-chain tackle, in which case a pair of Lewis bolts or 'clams' (Pl. 28 c) are inserted in the holes in the stone. The tackle should be operated from behind the stone as it rises, and from over the bed-stone as it is laid back on the floor, so that the stone cannot fall on the operator.

A millwright who undertook stone-dressing told me that after arranging to finish a job by a certain date, he and his men went to Norwich Fair; and returning late in their dog-cart they got so wet that they could not sleep (they intended to sleep in the mill). So at about three in the morning, they decided to get on with lifting a millstone;

but the noise, and the light they used, drew the attention of the police, who came up and banged on the door, demanding to know who they were, and what they were up to!

THE WORKING FACE OF THE MILLSTONE

Bed-stones are generally flat, but runner-stones are slightly concave on the working face, especially near the centre, to improve the feed and admit a little draught; grain thus progressively passes through a narrower and narrower space as it is worked from waist to skirt. Both upper and lower stones, which are mostly delivered with a smooth surface by the makers, have to be dressed with furrows and grooves, ten or eleven segments or harps usually being laid out for gristing and similar work.

The longest furrows are almost radial, and all of them are about $1\frac{1}{4}$ in. wide, and their purpose is to break the grains of corn. One edge of the furrow is cut as sharp and clean as possible; but for flour-milling the edge should be chamferred with a flat mill-bill to obviate cutting up the bran, as a Peak stone might do. The 'lands' between the furrows are finely grooved or feathered from six to twelve to the inch, like a file; and they skin the wheat kernel so that it can be ground. Indeed, there are reputed to be stone-dressers who can put sixteen stitches or drills to the inch; but others assert that such fine stitching is unnecessary, and is in fact gone inside of a week in grinding wheat (which is the purpose for which fine dressing is recommended). According to some millers, the stitching should normally be a little deeper towards the skirt.

To lay out a new clockwise (or right-hand) millstone, the perimeter is divided off into segments, usually nine, ten or eleven; and each segment is marked off from the skirt of the stone towards the leading side of the central hole, so that the marking runs at a tangent. A chord is then drawn across between the outer tips of two adjacent segments; and on the right-hand segment, as the miller faces the stone, the chord is divided into eight portions which are then ruled off parallel to the left-hand boundary of the segment. The resulting little triangle at the right-hand corner is a furrow, as also are alternate parallel divisions from right to left. The four intervening strips will be the lands on which the stitching is to be put in, and the furrows will be cut with a square edge at the left side and slanted off to register with the land on the right, so that each furrow is like a narrow valley with a ridge on the left.

The upper and nether stones being dressed alike, the grain is brushed into these ridges as it grinds between the lands, and is cut by the opposing square edges rushing against one another with a scissors action, which simultaneously guides the grain and meal down the furrows until it emerges at the 'skirt' or periphery into the stone case. From there it is swept round into a shoot or spout by the draught and friction, or nowadays by a brush or a metal 'paddle' or 'sweeper' fixed to the wall of the runner stone. Most millstones revolve with the sun, i.e. clockwise when viewed from above, but if they run the other way all the grooving is reversed, on *both* stones.

Millers prefer clockwise (or right-hand) stones, as explained in Chapter 4, p. 62, 'Rotation of Sails'. In Pl. 28*b* the miller is having to dress a stone left-handed in Syleham Mill. This has anti-clock sails with head and tail wheels each driving one pair of anti-clockwise stones, since both driving wheels have the cog-ring facing rearwards. (Some post-mills with anti-clock sails and two pairs of clockwise stones in the head have the tail cog-ring on the forward face of the tail-wheel so that it drives its stone clockwise, or 'with the sun'.) Clockwise spur-gear mills (with anti-clockwise stones) are found mainly in neighbourhoods where the finer points of windmill design and operation were not well understood.

It is easier to dress the upturned runner-stone than the bed-stone, because the bed-stone lies nearly flush with the floor. When two men dress a pair of stones, the senior takes the runner-stone while the junior lies on his side (or kneels, if space is limited) to deal with the bed-stone.

STONE-DRESSING

Millstones are dressed or 'cracked' mainly with a mill-bill—a specially tempered steel pick set in a wooden stock or handle called a 'thrift'; and the tap-tap-tap of the stone-dresser's bill was once a familiar sound in the quiet countryside, audible to passers-by if the mill were near the roadway.

A pair of stones provided a very full day's work for the miller and his mate—dismantling, dressing and re-assembling ready for grinding; in fact, two pairs would usually take up three days, and inexperienced men might even spend three or four days on one pair of badly worn stones. A single-handed travelling stone-dresser might take three days at 15s. per eight-hour day to do one pair. These semi-itinerant workers, who were among the individualist characters of the countryside years ago, may still be recognized by the little black bits of steel, rather larger than a pin's head, that are liberally embedded about their hands and sometimes in their faces—chips from their highly tempered mill-bills. These specks of metal drive right under the skin where they cannot be removed, sometimes even reaching the bone, but some of them eventually work out. Spectacles will protect the eyes as the particles will not break them, though they will scratch them. To protect the knuckles a leather or felt shield can be fixed on the thrift, or a thick glove may be worn.

The stone-dresser will usually have two or three spare thrifts with him, because they sometimes split, and also a pot of ochre or black for taking levels on the stone.

Skilled Work

In dealing with a 'kind' or 'open' stone—that is to say, one that has not a close or fine grain—it will be inadvisable to insert a very prominent 'dressing'. The tendency with such stones is to pass the grain through unground, and further inequalities put in with the aid of a 'pritchell' might aggravate matters so that whole grains are found amongst the meal. Nevertheless, a dozen mill-bills might possibly be blunted in dressing the furrows, at a present cost of about 28s. a dozen for re-tempering; but a single pritchell, shaped like a pointed cold chisel, would do the whole job.

'Open' stones become very irregular in general level of surface. Then the 'staff' or 'wood-proof'—a long dead-straight and heavy mahogany bar smeared with red ochre—will be put over the stone in all directions, and it may be found that in consequence of the uneven texture there are some unworn high spots that can be chipped away with the mill-bill (or a mill-pick—a pointed bill used in the thrift, and favoured by some millers for the preliminary levelling) before proceeding with the 'dressing' (see Pl. 28 *b*).

Mechanical stone-dressers *have* been invented, but I have never met any miller who found them effective.

There should be in the mill a metal 'prover', carefully protected in a case, for proving the 'staff' before use.

A few millers believe that, especially for grist work, it is satisfactory to leave both stones dead level and not take the extra time and trouble of getting the runner-stone a trifle concave; but this flattens, instead of grinding, the grain.

A typical farmer-miller, whose watermill did only grist—no flour milling—said that the old school to which he belonged believed that barley meal should be fine and soft; consequently they did not dress their furrows too deep, as the grain would then feed through too quickly and easily. Moreover, they put only about three drills in the longest 'land' in each segment, and a couple in each of the other 'lands', starting them from the eye, and running them parallel to the furrows, but reaching only half-way out towards the periphery.

Working round a clockwise stone little by little in an anti-clockwise direction, the miller will select twelve or fifteen of the fifty or more mill-bills kept in a busy mill, for they only remain really sharp for 10–15 min. work, and he will usually do the stitching or cracking first, while they are keen; and the furrows later. Beginning on the 'land' nearest in front of him, at the periphery of the stone, the miller chips at the rate of perhaps 120 strokes a minute, with frequent pauses, making a set of little grooves parallel to the furrows and as wide as the 'bill', till he has got probably a dozen grooves across the width of the 'land', then the process is repeated a bit farther up the 'land' and so on till he approaches the centre of the stone.

Levelling the Stones

Before replacing the runner-stone it is necessary to check whether the head of the stone-spindle protruding up through the bed-stone revolves dead true to $\frac{1}{32}$ in. in relation to this stone. Otherwise the runner-stone will lop from side to side and grind on the lower one, with a consequent risk of sparks and fire, while whole grains will emerge one side, and hot over-ground meal on the other.

To safeguard against this a wooden trammel or 'tracer-bar' (see Pl. 28c), 2 ft. long, with a square eye exactly fitting the stone-spindle, is placed over it, a separate tracer or 'jack-stick' usually being necessary for each spindle owing to variations in dimensions. Near the outer end of this bar is an upright quill—a chicken's feather —wedged through a small hole, with the stub coming exactly in contact with the bed-stone (which has first been trued with a mason's level or a 'gable') and a thumb-screw slightly opens and closes a little slit in the bar to secure fine adjustment of the quill. While the stone-dresser watches the quill minutely, someone downstairs slowly revolves the spindle, as the bar would be thrown out of alignment if it were itself directly revolved. A single-handed miller may, however, turn the neck of the spindle with an iron 'grip'. Should the spindle prove to be out of true it will be corrected by adjusting the four screws in the bridging box until perfect accuracy is attained—a process called 'brigging the spindle'.

After repeated dressings, a burr-stone wears down so much that the first retaining or binding ring has to be removed, and later perhaps another one will have to go, after having been meanwhile forced down against the third ring with cold chisel or drift, and club hammer, to give a little more clearance. As a rule, if the upper stone is (say) 3 in. below the top of the stone-case or vat, there will be plenty of burr left. If 6 or 8 in. down, there is probably no more than an inch or so of burr and the rest plaster of Paris, so the stone will not be worth much to a prospective purchaser of the windmill, and this is a consideration, as millstones were advertised at over £30 apiece even in bygone days.[1] Great care was taken to see that the surfaces did not make contact with each other and become damaged when being replaced after dressing, a liberal bedding of bran being distributed over the nether stone for this reason. They ought never

---

[1] One miller told the author that he could not say what price new millstones fetched in his day, because they were beyond the means of a country Mill doing grist at 1s. or 1s. 3d. a sack, and he had to struggle along with second-hand ones.

to be put down on a damp day, because there will be condensation on them and they will not grind; but stones do not sweat once they are face to face.

In a nineteenth-century history of the Scottish people, it says that a hundred years before, the tenants of an estate were expected, as a part of their free services to the laird, to take home the millstones of his mill at their own expense, and re-dress them. The stones were taken by rolling them on edge with a pole passed through the 'eye', and steadied by two or three crofters on either side. They trundled the stone themselves, or hitched it to a horse with a trace to either end of the pole.

The Mill-bill

Although, in the heyday of windmilling, a flourishing business would have several dozen mill-bills laid out ready for the travelling stone-dresser, the visit of this roving expert to the smaller country mill was often a not very enjoyable time for the mill boy. He might be kept more busy than he cared to be, turning the handle of the grindstone from morning till night to keep the miller's half-dozen serviceable bills keen-edged enough for the stone man to press on with his work. Even so, there came a time when the hardened casing of the tip had been entirely ground off, and the bill must be 'hardened out' again.

So important is the skilled job of tempering the bills that it has actually been hinted that the use of millstones in windmills and watermills may die out for want of the skilled knowledge of just how to temper the bill so that it will do its job.

Tempering was done by heating the bill to just short of cherry-red and plunging it into cold soft water—rain water, in fact—and leaving till cool; but various smiths recommend different methods, some preferring to give the final douche in oil instead of water. Others prefer only to plunge the 'bill' momentarily in water when cherry-red, then watch the colour run back down the point till it is straw-coloured, then dip it right in and leave it to cool. Mr Horace Barker of Norton, who worked several windmills in his time and dressed the stones at Pakenham Mill (see Pl. 28c), says that in tempering a mill-bill you *must* plunge it straight into *salt* water, at exactly the right moment; otherwise the bill will either be soft and useless or it will chip and fly to pieces. If cooled as quickly as possible at the right moment the bill will be tempered well through, especially in the last half-inch. To re-harden it superficially, at the mill, it may be heated up red-hot and rasped with a rough file and then plunged. Often in the past a man collected mill-bills from several mills and paid a blacksmith a fee for the use of his forge for a day; and a stone-dresser would sometimes temper all the bills himself at the blacksmith's for the particular mill where he was about to work, because the average blacksmith had not always got the knack.

FLOUR-MILLING

Flour-milling is a big subject; and many of the following notes are based on William Halliwell's *Technics of Flourmilling* (written in the twilight of windmilling), supplemented by the late Hector Stone's rather discursive comments and explanations.

In windmills and watermills every particle of the milled wheat from the stones went into the dresser; and the 'throughs' (the materials that passed through the wire gauze) were the flour of commerce of last century. The rest was an impure kind of bran, because no single machine of that era separated pure flour from pure bran even if another machine removed the residue.

Wheat yields about 70% pure flour, with 14% 'pollards' or 'thirds', $14\frac{1}{2}$% bran, and $1\frac{1}{2}$% dust or lost material. Pollards, thirds and bran are sometimes loosely referred to as 'offals'.

The wheat berry—the grain that goes to the millstones—consists of 84% kernel, $14\frac{1}{2}$% bran, and the remaining $1\frac{1}{2}$% is the germ from which springs new life if the grain is sown.[1] This germ, situated at one end of the grain, against the stalk, has a pronounced yellow colour and gives off the pleasing nutty smell characteristic of wheat. The bran is the external covering of the kernel and is about $\frac{1}{250}$ in. thick.

[1] It was generally believed that the natural vitamins are in the germ, but it is now known they are in a membrane between the germ and the endosperm (the starchy interior).

WINDMILL FLOUR

The windmiller has passed his wheat-meal (which would make wholemeal bread, if used as it came off the millstones) through a bolting machine that grades the 'stock'—a general term for whatever is fed to a machine—into three bins or sacks through three different silk meshes, the residue that has not dressed through the silks being largely bran, because it comes off in sizeable flakes in the millstones.

Nowadays the wheat is separated from extraneous matter by being washed, cleaned, brushed, dried, graded and conditioned (to make 'hard' foreign wheats millable) by machines that neatly remove large seeds, small seeds, stones, nuts and bolts, chaff, dirt, fibre, 'smut' balls, barley and everything else that a robot could be expected to identify and extract. Even discoloured grains can now be selected and removed by an electronic machine.

The washing process alone, which is of great importance, could never have been arranged within the limited confines of a windmill. If water had been available, there was no power to drive the cleaning machines; and foreign wheats are sometimes so hard and dirty that washing is an absolute necessity.

When wheats have to be mixed without washing, as in a windmill, they should stand for several days after mixing, to communicate the moisture of the softer grains to the harder and drier ones, as no milling process can be properly performed on an uneven sample, because the setting of the machine or millstones to achieve the best results would necessarily differ from time to time.

Before grinding, the wheat could be divested of dust, sorted into small and large grains, and freed of stones and grit, in a horizontal worm-feed 'separator' made of a series of wire meshes built into a very slowly revolving sleeve, with the usual range of funnels and receptacles underneath; but it ought to be realized that even when such machines were invented, a little old post-mill was limited in its power and in its accommodation, and many a miller contented himself with installing the solitary wheat-meal dresser in the tail as seen in Fig. 3.

Another innovation was the 'screener', to be used after, or alternatively to, the separator. It resembled a dresser with one grade of wire only and ran at a similar speed, and its speciality was brushing the dust from the grain. Then came a 'smutter' for such grain as required it; this was a vertical cylinder in which high-speed revolving beaters (450 revolutions a minute) thrashed the gravitated wheat clean of the obnoxious 'smut'.

Then came the milling process, followed by the dressing in the 'wire machine' which would have a 120-mesh wire for finest flour, but more usually 70, 50 and 30 meshes for firsts, seconds and thirds. Alternatively, the bolting machine made fine flour, besides having the advantage already mentioned of handling hot meal from the millstones and delivering it cool enough for bagging off. A century ago, in a bolter with several grades of silk, the stock was separated into first-grade flour, from which 'wheaten' bread was made; second grade for 'standard wheaten' or wholemeal flour; supers or middlings for coarse 'household' bread (but mainly for pig and cattle food); and bran. Another account, written in 1841, said that wheat was ground into fine flour, seconds, fine middlings, coarse flour, bran, 'twenty-penny' and pollards. Middlings are a starchy food somewhat superior to the stock adhering to imperfectly dressed bran; pollards or thirds are a lower grade mixture generally containing meal and bran.

In a windmill, 'sharps', often spoken of by millers, are a grade of offal superior to 'toppings'.[1] According to the late Hector Stone of Upminster, flour dust is extracted from them and the granular residue sold as 'semolina'; and he further explained that semolinas are little hard grains or particles of wheat like coarse sand, dislodged from the berry in scraping the bran flakes.

Grinding, Output and Tolls

Gristing is somewhat monotonous once the newly dressed stones are regulated: barley, which should make *meal*, not rough husky stuff,

[1] Sharps in the west country are the same as 'thirds' (the third cut off the dressing machine) elsewhere, and 'toppings' in the south midlands. In different districts superfine toppings are called 'superfine middlings', and so on.

grinds very dusty, and some millers subject to asthma lay a cloth over the mouth of the sack and even wear a small respirator when the dust flies. Barley grinds warmer than oats because the stones are screwed down harder to grind finer, and it is also said to retain the heat more. If it is damp, it may clog in the hopper, which would precipitate a disaster if it stopped feeding into the stones; or it may clog the furrows so that the stones do not grind.

When wheatstones are nicely adjusted and running well, they 'sing' pleasantly so that they can be heard across the mill yard.

For cleanliness, experienced millers prefer an 'overdrift' mill (as, indeed, windmills should be, for lightness and efficiency of driving gears), because underdrift gearing becomes very dirty and clogged with a compound of meal dust and grease, besides being generally inaccessible for repairs and adjustments.

Modern flour will usually take up to 50% water, in the mixing process, and as but a fraction of this is evaporated in the baking, Sir Richard Phillips's statement a century ago that 2 lb. of wheat made 3 lb. of bread still holds good. Fifty years ago a milling text-book stated that when a sack of flour (280 lb.) was to be baked, it was mixed with 1¼–2 lb. of yeast, 60 or 70 quarts of water (300–350 lb.) and 3–3½ lb. salt. It had to stand 6 or 7 hr. to ferment or sponge, and then be placed in an oven at 500° F.; but the probable evaporation of water was not mentioned. The late Mr Benjamin Burton—foreman of Slough steam-mills for 42 years—said that 400 lb. of English wheat produced 84 'quartern' (4 lb.) loaves, and 400 lb. of foreign wheat or 280 lb. of flour, 96–100 'quartern' loaves, according to the moisture content of the wheat. It is illegal to sell by the true quartern, which is 4 lb. 5½ oz.; and it may be of interest to add that a peck loaf weighed 17½ lb., although a peck of flour was 14 lb.

Millers' statements as to the number of sacks, bushels or coombs of corn they have ground are often confusing, and the following may help to light up their remarks. One 'sack' equals four bushels or one coomb, and two coombs equal one 'quarter'; but the avoirdupois weight of a bushel is not constant, thus a bushel of wheat is 63 lb., barley 56 lb., oats 42 lb. and beans 77 lb; so a coomb of wheat is 252 lb. or 18 stone, barley 224 lb. (16 stone), oats 168 lb. (12 stone) and beans 308 lb. (22 stone), and a quarter of wheat is 4½ cwt., barley 4 cwt., oats 3 cwt. and beans 5½ cwt.[1] The sack of grain is called a '4-bushel sack'; the term '20-stone sack' (2½ cwt.) is applied to a sack of flour; the coomb is a measure of grain, not of flour.

A hundred and fifty years ago, when a census of windmill and watermill output was called for by the High Sheriffs (during the 'invasion' scare of the 1790's) corn was also measured in loads— usually the standard 'horse-load' of 2400 lb.; although sometimes the 'man-load' of 300 lb. was indicated (approximating to the 'porter's load' which is 20 stone); also the 'bag' which is, and probably always was, 3 bushels or three-quarters of a coomb (*The Posse Comitatus*, by John Penn, High Sheriff of Buckinghamshire, 1798).

The census figures indicated further that 3 units of wheat were considered to produce approximately 2 units of *flour*; and an old Lincolnshire miller said that in his day, the 18-stone sack of wheat produced 13 stone of flour (72% extraction) and 5 stone of offals. The whole of the offals was taken as 'toll', and no account was rendered either way. He thought that in earlier times they took 2 stone of flour, and delivered 11 stone of flour and 5 stone of offals.

When grinding barley, the miller fetched it from the farmer, ground it, and delivered it 6 or 8 miles, for 2s. per quarter (or 3s. by steam) with (to the sack) 4 lb. waste, which fell on the floor or stayed in the stone-cases; and this they kept for pig-feed.

The owners of twenty-five windmills in Buckinghamshire estimated in the 1798 census that they could grind approximately 400 quarters of corn per week between them; but their individual figures varied between 2 quarters (presumably meaning on an average during the season from harvest until the calm of the following summer), and 30 quarters (which probably related to a good week in wintertime). A windmill converting 2 sacks of wheat into flour per hour might be developing 30 h.p. A roller steam-mill was supposed to take 10 or 12 h.p. per sack per hour, with lighter and more fragile machinery; but the product was not as good, although cleaned and purified.

---

[1] A 'quarter', avoirdupois, is a quarter of a hundredweight—28 lb. Beans are 4¾ cwt.—532 lb.—in modern milling.

In a south-westerly gale on 17 October 1949, Mr Reynolds Wright of Friston post-mill, Saxmundham (see Pl. 1), ground 40 coomb of *grist* (about 3½ tons) in the tail-stones with the single surviving pair of sails; and he said this was about as much as she had ground the previous 12 months! Friston Mill was the largest and most powerful post-mill remaining in service, until Cross-in-Hand Mill, Sussex was recently restored to work; and Friston measures some 50 ft. to the roof ridge. Her 80 ft. patents (9 ft. wide) are carried on a fine old 66 ft. stock of 14 × 12 in. section at centre, and with her full complement of sails, as well as wind-shaft and head and tail wheels, Mr Wright estimated she carried 8 tons weight on the weather-beam. On this memorable occasion he set her in full sail and put through all the corn he could, running on late into the evening, just to see what she would do when she was set in a good wind.

Old neglected stones, with little dressing left on their faces, will still grind a little grist after a fashion—perhaps half a coomb an hour, but heating and spoiling the meal in the process; for there is an art in keeping stones so dressed and adjusted that they grind coolly and efficiently.

Sussex ground oats, it is said, can be ground on Peak stones properly furrowed, but without any 'cracking', because the 'gritty' grain of such stones presses the bran without cutting it, and at the same time removes the flour meal. Freshly re-dressed stones should do up to 3 coomb of grist an hour when the wind blows nicely; but flour-milling, although more profitable, was less speedy, with the dressing to do as well as grinding. Even in the old days of very long hours it was a notable feat to grind and dress 100 coomb (2¼ cwt. sacks) of wheat in a week, and 4 bushels an hour was held to be a good average; but the giant Southtown Mill, Gorleston, with four pairs of stones (all on the seventh floor) would grind from 8 to 10 coomb of grain per hour. It is interesting that by working 80 hr. in the week, as she might well do in winter, and taking 1s. 3d. per sack, this mill could get through £50 worth of grinding. She was a Suffolk mill; Gorleston was transferred to Norfolk after she was demolished.

Wheat can be ground too fine, and the only way to achieve pro-
ficiency is to learn from an experienced miller. The finished wheat-
flour should have a sort of yellowy sheen about it when inspected in
a good light (more obvious if the sample is placed on a board and
slightly moistened). It should normally fall apart as if slightly
granulated and not tend to hold together in clots; but this tendency
is also influenced by the moisture content and the softness of the
wheat. A few dark specks may not matter, but there should not be
patches of discoloration; and some millers find that in comparing
two samples of flour in an even light one's first impression is usually
the right one, and doubt will enter the mind if the samples are studied
too long and minutely.

Smuts, mustiness or sourness are readily detected by smell, but the
final test, if still in doubt, is baking.

In the Middle Ages, the miller took toll of 4 lb. of meal from every
coomb (using the 'strike' or 'strickle'—a wooden staff—to level off
the meal in the measuring bowl), and fattened pigs with it, but in this
mercenary age, and indeed for many generations, cash transactions
have been preferred, as witness the following notice posted in
Darsham Mill the year it was re-erected from a neighbouring site:

DARSHAM WINDMILL

Notice is hereby given

That from the 6th day July 1809, whoever has any corn ground at
this Mill shall pay the price of 1s. per comb for grinding and 1s. per
comb for dressing Wheat: 1s. 6d. per comb for grinding Hog Corn,
and the accustomed for waste in grinding and dressing of same. The
above prices to be paid if demanded before the Corn is taken out of
the Mill, otherwise a part thereof shall then and there be left, equivalent to the amount of the charge for grinding, and the quantity to be
left shall be ascertained by the then current price of Corn of the like
description and quality.

Chapter VIII

A MILLER'S LIFE

❧

BEFORE describing present-day experiences of windmilling, let us read over what James Edwin Saunders had to say in his reminiscences of working a decrepit old post-mill at Stone (Buckinghamshire) at the age of eighteen. True, his mill was not so old as he believed—probably under 200 years at that time—but it had been moved, which usually weakens the structure.

'If anyone wants to know what it is to rough it', he said in *Reflections and Rhymes of an Old Miller* (1935), 'let him try driving an old-fashioned windmill day and night, and in the condition my first mill was in. A modern windmill is one thing, an antiquated post-mill 400 years old, as mine was said to be, is quite another.

'Many a time when I was out shifting the cloths in a storm, the water has run off them down my arms and out at my trouser legs. Of course there was no chance of getting dry clothes until I went home, and in the winter they have sometimes been frozen on me for hours....

'I can only once remember being frightened by a storm at the old windmill;...It was in October or November, at a time when I had been so busy that I had not kept a proper look out for storms, which we always reckoned to do in wild weather; and a tremendous hurricane caught me unawares.

'My first warning was that the mill was running faster and faster, but I was not really disturbed then until I had put the brake on and gone down to take some cloth off. Outside it was as black as pitch. I felt my way round to one sail and was just beginning to uncloth when the gale came on like mad. It blew me against the round-house, and away went the sails as if there was no brake on at all. I shall never forget how I rushed back up the ladder. The whole mill rocked so that the sacks of meal that were standing in the breast were thrown down like paper, but I got to the brake lever somehow and threw all my weight on it, but it hardly seemed to check her. I knew that if the brake was kept on she was bound to catch fire, so I let her off, and round she went, running at such a rate that the corn flew over the top and smoke blinded and suffocated me.

'Then I *was* frightened. I expected her to crash over at any moment, and, half overcome by smoke and din, I dashed down the ladder and out into the close. But no sooner had I got there than I thought it was cowardly to run away, so I went back to have another try to stop her....By this time the light had gone out, but I got up into the stone floor and once more put the brake on, this time jumping up on to the lever and putting my shoulders under the next floor so that I could prise it down with all my strength. The sparks were flying out all round the brake as she groaned and creaked with the strain, but it still didn't stop the sails; and I doubt whether anything could, had not the hurricane itself subsided as suddenly as it sprang up.

'In that brief period I had one of the strangest and most vivid experiences of my life. As I clung to the brake I heard a crash in two places and I knew that two of the timbers had snapped. I felt certain she was going over, but there was no chance of escape and I kept my position....How long I hung there I cannot tell, probably not more than a second or two, but in that short interval the whole of my life flashed before me with a vividness indescribable and unforgetable.'

MEMORABLE TIME

This memorable but by no means unique experience happened in the early sixties; but it might have happened many a time in the 'good old days'. Here are some of them! 1315—wettest summer in history, when corn rotted as it stood in the fields; 1684—the Great Frost that held for 13 weeks; 1703—a terrible wind, highest ever known, on 26–7 November, when Downland millers in Sussex were struck by billows of sea water hurtling through the air, and countless windmills were blown down or wrecked; 1716—Great Fair on the Thames; February 1760—a 'most terrible storm' that set windmills ablaze; 1762—Great Snow for 18 days; 1839—terrific and destructive gale; 2 August 1879—terrible thunderstorm and cyclone (several tower-mills decapitated); 'Black Tuesday', 18 January 1881—worst blizzard in living memory, innumerable windmills seriously damaged and several blown down; and finally the destructive blizzard of 28 March 1916, rendering further repair to a great many windmills unjustifiable.

July 1947 brought a freak cyclone to Mr Jack Penton's post-mill at Syleham, on the Norfolk–Suffolk border. Coming up from astern whilst the mill faced a light east wind, the hurricane caught the mill full on the tail before she could possibly be luffed with the hand-crank—no time to fetch spanners and uncouple the fly, and cranking against the fly is hard and slow work. Before she was a quarter round, the cyclone was in full blast, with a blinding rainstorm of such intensity that the windmill was invisible at 100 yards in daylight. Notwithstanding the miller's efforts, she would not come round quickly as she should have done with the gale driving the fly; and in less time than it takes to tell, the wind-shaft, urged on by the furiously revolving sails, tail-winded as they were, had torn up the wooden 'keep' on the tail-bearing and was thrashing around the upper floor, smashing bins in its path and tipping forwards so that the sails bashed the round-house wall and were damaged; then just as the sails threatened to part company from the mill, the storm abated, as at Stone, and all was calm. Having been unable to luff the windmill, the miller was helpless and could not have saved her had the hurricane persisted.

The late Mr J. B. Geater (Pl. 26b), writing to me at the age of 94, said that when his post-mill was struck with all the cloth on, she 'went off like a rocket and the sails spun like a top'. One cloth was torn off and blown across the fields and another split from top to bottom. This eased her up and saved her.

WIND FORCE

From Mr Jack Penton's experience it can be seen that wind in force was a matter of the most serious concern to every windmiller, much as he may have welcomed the wind in moderation.

The maximum wind pressure to be recorded in an average storm is unlikely to exceed 25 lb. per square foot, which is equivalent to a tonnage evenly spread over the complete frames (with cloths furled) of a common-sailer amounting to about their own weight, and the same would apply to most sails. The subject has been scientifically investigated, and it is found that gusts of 72 m.p.h. are not often exceeded in England; they never exceed 90 in the windmilling counties —only in wet areas where corn is not grown. Wind comes in gusts and lumps, influenced by ground undulations, buildings and trees, and upper air currents; and eddies and downward currents are similarly produced. Records taken at Bidston, Cheshire, in the 1870's showed that in gales of maximum velocities (of from 60 to 90 m.p.h.) pressure *in the heaviest gusts* ranged from 40 to 80 lb. per square foot.

Wind velocity is greatest at the top of a mill, whether in flat country or not; and at heights to which it is impracticable to construct a windmill economically a sufficient wind force to drive it would *always* be available.

The writer's experience in heavy weather at Brill is that an otherwise steady wind (not gusty) has a sort of periodicity, so that it beats rhythmically on the mill body. Wind is accelerated as it mounts a big hill from a level, hence very high velocities occur at Brill;[1] and in February 1949 the mill sails made several revolutions with *bare frames and the brake on* in a terrific head wind from the west (not tail-winded), yet there is only a very moderate 'weather' on these sails.

As far back as 1841, Sir Richard Phillips (*Million of Facts*) had a fair estimate of wind pressure, and he correctly believed that the pressure always quadrupled as the wind speed doubled—thus the pressure was about 0·8 lb. at 12½ m.p.h.; 3⅛ lb. at 25 m.p.h.; 12½ lb. at 50 m.p.h.; 50 lb. at 100 m.p.h.; and 200 lb. at 200 m.p.h. This represents 1 lb. at 15 m.p.h.; plus another 2 lb. at 25 m.p.h.; plus another 3 lb. at 35 m.p.h. (i.e. 6 lb.); and 10, 15, 21, 28 and 36 lb. at

[1] Unfortunately no records are taken near any windmill overlooking a steep declivity, such as Brill. The Bidston recording site seems most nearly comparable with Brill, and gusts from 77 to 91 m.p.h. are registered there most years.

45, 55, 65, 75 and 85 m.p.h. and so on. Another writer gives 0·5 lb. at 10 m.p.h. (brisk), 2 lb. at 20 m.p.h. (very brisk), 12¼ lb. at 50 m.p.h. (very high wind), 27¾ lb. at 75 m.p.h. (hurricane) and 50 lb. at 100 m.p.h. (tornado).

TAIL-WINDED MILLS

Occasionally it has happened that the sails and wind-shaft (and the cap also unless the tail-bearing broke away) have been tipped so far forward on a tail-winded smock-mill or tower-mill of squat design that, although not actually hurled to the ground, they have lodged with the wind-shaft laid over the neck bearing, and the two lower sails set astride the mill like a pair of sheer-legs, with their tips dug into the soil, whilst the forward centering wheels of the cap have probably jammed behind the mill-curb. If the cap has lifted, the sprattle-beam may have parted company with the top gudgeon pin of the upright shaft. In such cases the sails have often been restored to their rightful position without dismantling.

The first thing to do is to see what security there is to prevent the sails toppling right over when disturbed; and if the centering wheels have got clear of the curb, the brake-wheel will probably be fouling the curb; but if the cap has not lifted much, the wind-shaft may have broken the 'keep' at the tail, and the brake-wheel will be caught up against the weather-beam.

It is necessary to lash the wind-shaft and brake-wheel firmly to the upright shaft below the wallower, after laying a strong baulk on the top floor from the shaft to the 'front' wall (under the weather-beam) to counter the tension on the shaft. Strap two big poles together so that when hauled up and manoeuvred into position they will form a pair of sheer-legs crossed at the top to make a fork, which will fit between inner and outer stocks to support the canister. The props will need wide boards under them (and firmly secured to them) in a wet winter, to prevent them sinking. The canister may then be gradually lifted by levering the sheer-legs up to their job with pinch-bars. Meanwhile a couple of men will have the strongest available worm-and-chain gear coupled from the tail of the wind-shaft to a lower floor-beam—one operating the gear and the other watching from the top of the mill to maintain contact and relay signals and instructions. Having coaxed the tail of the shaft down, the shaft may need pulling farther back over the sill, and if the neck-wear is still *in situ*, the sails can be lifted a little more in order to remove it, until the shaft is correctly placed, when it can again be raised a little (by small sheer-legs used in the cap) and the bearing replaced. Even if not actually damaged, the brake will probably have been displaced and will need reinstating; any other damage, which might be quite extensive, will be duly appraised and made good.

THE FANTAIL

Except during a thunderstorm, fantail mills are not likely to become tail-winded if the fan is well designed and geared, free-running and well set up, with plenty of black grease on the cogs, but they are generally a bit slow to follow a change of wind in a 'small' breeze. Should the fan spin round and fail to wind the mill, owing to a broken cog on a tower-mill or smock-mill curb, a loose truss of straw should be thrust into the fly (not a modern bail) to choke it. The fly must then be roped up until repairs can be executed to the cog-ring by fixing new cogs with set-screws.

Moreover, in such cases, and also in the event of the fly itself being broken up in a storm, precautions must at once be taken to prevent 'spring' or 'patent' sails being damaged by a tail-wind. Extra vigilance is called for from the miller to watch the wind and keep his mill or mill-cap hand-cranked into the wind; and he should fix his sail shutters open at night. Mr Reynolds Wright, of Friston post-mill, Saxmundham (Pl. 1), says that if the fly on his mill is only 6 ft. out of a tail-wind—say 5 degrees—on the tramway, she will easily pull right round into the wind very quickly, with no need for anxiety at all.

STORMY WEATHER

Experienced East Anglian millers say that the following procedure is effective in combating a cyclone with a fantail mill, providing they have sufficient warning of its approach. Tie up the fly and unbolt it in order to luff the mill by hand a quarter out of the wind. When the storm is too near for the mill to return to its former position (back to the oncoming storm) couple-up the fly and release it. She should then be caught up by the contrary wind and turned into it.

If the mill has actually become tail-winded in the miller's absence, and the sails have started to run away, much the same will apply (at any rate in theory!)—the fly will not be turning much, for she will be dead in line against the wind. She can then be tied and uncoupled, and if the sails can be stopped, lash the stock to the nearest anchorage, such as a tree (the 'whip' will not stand it in a rough storm). Wind the mill round by hand, taking care to pay out or haul in the sail rope in order not to pull on it, and re-set the fantail; after which all should be well; but it goes without saying that theory works better on paper than in a hurricane.

There is a story of a millwright who was called in to stop a windmill (kept by a publican) in a great storm, and knowing that she would hold on for a bit he blandly remarked, 'I shall require a quantity of oil before I can stop her!' Whereupon the miller brought forth an oil drum from his outbuildings. 'No, not that sort of oil', said the millwright, 'I want it out of a barrel!'

SAFETY PRECAUTIONS

Many a windmill, particularly a common-sailed tower-mill, is not blessed with a really effective brake, and the addition of multiplying pulleys for the brake-rope, or of several heavy weights on the brake-lever, is not unusual—even the big patent-sailed smock-mill at Upminster requires this addition, or else the sails can actually be moved with a boat-hook while the brake is on.

If the brake is poor a common practice with post-mills to stop the sails revolving is to leave the stones engaged and the sail shutters open at night. Should the wind change, the mill and its equipment will then turn bodily. Do this with a smock-mill or tower-mill, however, as a forgetful miller has sometimes done when acquiring such a mill in place of a post-mill, and something has to go, because the fan is trying to turn the upright shaft as well as the cap, which it cannot do with the millstones engaged. So, either it tears the cogs out of wallower or curb, or else twists the upright shaft, if this should prove to be made of wood of insufficient strength.

If it is not desired to let the sails run free during the night, the brake-wheel should always be secured with a stout chain hooked round the sprattle-beam or some other strong anchorage. A very stout log, preferably of oak, serves the purpose in some mills, but it is not always possible so to place a log through the wheel that both ends are secure in the event of it revolving either forwards or backwards and one way or the other the wheel may possibly throw the log out of position.

Every care should be taken to detect weak sails or stocks in good time; and the retaining straps and wedges should be periodically tightened. Loose or loosening sails often betray themselves by a jerky motion as they rise and fall in their great orbit through the air, and a 'bump' can be felt within the mill.

THUNDERSTORMS

Precautions against lightning seem to be a matter of opinion—the iron sail rods of shuttered sails are generally said to attract lightning; and if they are 'patents' the electric current may pass down the iron striking rod and round the inside of the mill to the sack-chain, which has sometimes been known to be welded into a solid stick so that it has had to be cut up and scrapped.[1] Thence the lightning runs to the ground, perhaps splitting some timbers or brickwork on its way.

It is commonly believed that the danger of being struck is made less by setting the sails on the cross and winding all the chain on to the sack bollard to break the direct path of metal to the ground. It is a fallacy to suppose, as is sometimes said, that lightning never

---

[1] One miller was killed by lightning whilst handling the sack-chain.

strikes twice in the same spot. Several windmills have been twice struck; so have several church steeples.

Anyhow, it is not thought that a lightning conductor is much use; the sails are bound to project higher than the conductor; and in Holland the new steel or alloy sails do not seem to have led to disaster, as might perhaps have been feared; some of the German metal-sailed mills have a thin lightning conductor rod projecting from the cap.

As lightning conductors are no good unless well earthed in the ground, it would be very difficult to arrange an efficient one on a post-mill (which turns bodily).

MILL FIRES

Occasionally a cloth-sailer would get thoroughly out of hand. Some millers would then pull the mill a quarter out of the wind, with a chance of getting the cloths torn off (if they dared risk the gale on the side of the mill), jamming the stones down meanwhile, to choke them by feeding them with the maximum amount of grain, and throwing brick dust or broken glass into the brake to get a grip before the increasing volume of smoke turned to flames. On a black night with the sails racing round in a storm, all this is easier said than done; and should the miller be short of grain, one of his boys would be sent off as quick as he could go to beg the farmers round about to send up their grain immediately; they would usually do this, so giving the harassed miller a chance to try and ride out the storm under full sail.

Running out of grain before the sails could be stopped meant almost certain fire; for there is no escape if the stones meet and throw out sparks.

Everything about the mill is inflammable—the old dry wood, the flour dust, the sacks, the paint, the grease, everything—and, once started, the mill will quickly burn down, with sails hurtling round enveloped in flames—a most awe-inspiring sight—flinging pieces of burning wood in all directions, to the danger of the outbuildings and the thatched or wooden-walled cottage in which many a miller lived.

The only *safe* rule is to follow faithfully all the usual rules for fire-prevention in any kind of mill or workshop. Keep the mill clean and tidy; oil the bearings regularly; no smoking (a rule not observed by every miller and millwright); no candles; plenty of grain in hoppers in winter; no oily rags or ropes heaped in corners (these are liable to spontaneous combustion as well as being inflammable), and if any kind of water supply is available keep a stirrup pump or hose handy. If the mill is on a common, watch out for people lighting fires in windy weather. With all these precautions attended to, the miller has at least done his best to safeguard his inflammable structure.

GOOD AND BAD WINDMILLS—GOOD AND BAD FLOUR

During a visit to Friston Mill, when she repeatedly started and stopped in a moderate breeze (which Mr Wright estimated at 10 or 12 miles an hour, no more), I noticed that she restarted without a creak or groan, moving off as if the whole structure were firm as a rock. Old Wenhaston Mill, on the other hand, in a rough wind, rolled and swung this way and that like a ship at sea, so that on the top floor one almost had to hold on to keep one's feet. Wenhaston Mill was a tall, old, narrow-bodied mill, with disproportionately big and heavy sails. She had lost her fly, so that the wind sometimes struck her sideways on, unless the miller repeatedly luffed her with the hand-cranking tackle. Westleton and Saxtead Mills, of similar build with less ungainly sails, also rolled sharply from side to side, especially when accelerating, as each sail caught the full force of the wind aloft; indeed, it has been unusual in recent years to find a post-mill that drove steadily along without pitching about or creaking and cracking under the strain upon the timbers.

Movement is generally discernible in a smock-mill when she is working (and the cap can be felt to vibrate in a rough wind even while the mill is standing idle), but tower-mills are rigid unless very tall or slim; Southtown Mill, Gorleston, a tower-mill, is reputed to have swayed as it worked. Some post-mills, perhaps more especially those with fan tackle on the tail-pole, such as Cross-in-Hand (Sussex), luff into the wind with a noisy jerk as though a timber had broken; this is due to stiffness around the main-post, which prevents the structure from following round with the fly immediately.

Nevertheless, a little old post-mill would often get right into the work and romp through it at a fine pace all day (not being unduly cumbersome, especially with overdrift stones); and Saxtead Mill (which her one-time miller says is no longer 'go-able' because of her weak weather-beam and general old age) was an outstanding example of a mill that dug itself in to a good day's grinding. Very old cloth-sailers, too, were regularly to be seen running at a nice pace while neighbouring patent-sailed towers refused to budge; Quainton and Stone post-mills (amongst others of this kind in the Vale of Aylesbury) were good examples. One miller whose sails pulled well in a nice wind was wont to say that his sails 'do wholly draw out'!

WHAT OF THE FUTURE?

If a windmiller hopes to grind wheat for flour today ('stone-flour millers' they call themselves as distinct from the 'hog-miller' who grinds pig food—the only class of windmiller in business at the moment), one problem is growing a suitable strain of wheat to produce an attractive white flour. Red-skinned varieties, which big millers and farmers claim to be the only profitable strains in this country, do not readily grind white in millstones; but the late Mr Stone said that successful white-skinned wheats *are* grown in East Anglia (where windmillers are most likely to be interested), although rather costly to buy because they are not as *plump* as they should be for millstone grinding.

Ministry of Food requirements—lining flour bins with metal sheeting, removal of worm-eaten woodwork, and painting of interior walls to facilitate regular sweeping down—are also a source of additional expense to be considered, although perhaps not by any means premature from the point of view of hygiene.

Modern power-mill machinery makes a low-quality white flour, acceptable to the mass-production baker, out of almost any artificially 'conditioned' grain. White machine-made bread, in which the wheat germ has been damaged, and the vitamins removed, leaves only a starchy constipating mass. Real bread, as obtainable from windmill flour, is notable for its excellent flavour, good keeping qualities, and the fact that a smaller quantity of it needs to be eaten to sustain one's energy.

One of England's last windmillers—Mr John Bryant of Pakenham—solves the acute economic problem of his trade by loading his bins with 16 or 18 bushels of grain, setting the mill going and then betaking himself to the fields, where he ploughs and sows. He times his grinding by experience, watching his mill the while, and returning perhaps half an hour before he estimates the bins will need replenishment. Then he bags up all the meal, ground into a single bin, and puts another batch in hand. Grinding corn from his own extensive holding provides half the work for his mill, for many of his neighbours do their own grinding with a motor these days.

VISIT TO THE MILL

The windmilling enthusiast who is able to visit a windmill in full sail today is lucky; so few are left. Thirty years ago, in south-east England and East Anglia, many a mill turned throughout the year. Just before and after the second world war, visitors to the mill were welcomed—the miller understood an interest being taken in his lovely old 'standard', and welcomed the visitors.

Today, so rare and far apart are workable windmills that many casual visitors, with no lasting interest in the subject, are drawn from far and wide. They can become a great hindrance to the miller, already hard put to it to get through enough work to make ends meet. Millers begin to complain of lack of privacy, loss of time, and constant need for vigilance against meddlers and souvenir hunters, and one sympathizes strongly with these complaints; indeed, it seems almost as though the local authorities may have to undertake, if they will, to maintain an occasional windmill in action solely for instructional and educational purposes, even accepting a small loss on the working account just as they do with museums. The wind-miller cannot afford to provide the entertainment himself, much as he might wish to do so, not even at sixpence a time.

So before it is too late, shall we take a look inside an ancient post-mill, swinging along in a good breeze?

Creaking and swishing in the wind the sails sweep round majestically, with a gentle 'slap, slap' of the shutters as each sail comes down from aloft; for even 'patent' shutters swing to and fro a little as they alternately gather up the wind and discharge it. Over the broad and much-trodden steps, the fantail creaks and clangs a bit as the gears bite in a changing breeze to urge the steps imperceptibly around the tramway.

The wind is rising, and the miller seizes a weight from the reefing chain, thus lightening the load on the shutters and checking the gathering speed of the mill—he hasn't got his tenters set for a heavy wind at the moment, and the great runner-stone is already revolving with a rumbling, roaring note that fills the dusty air.

Big wooden cogs are meshing through their iron counterparts on the stone-nuts. The huge sails, accelerating again in a rising wind, travelling quite 30 miles an hour at the tips, throw the mill from side to side as they catch the wind aloft, bringing every corner of the old structure to life—fantail pulling harder too, so that the great body is wrenched about with loud protesting cracks, shuddering uncomfortably under the strain; chains rattle and sack-traps bang as sack after sack goes up to the top of the mill. The damsel, clattering rhythmically on the feed-shoe, rises to an incessant noise in the gale, so that the miller shouts to make himself heard; the bell alarm sings melodiously with that pleasant, continuous ringing tone set up by the vibrations that have, in fact, gripped the whole structure as the wind drives her furiously along; indeed, we involuntarily grasp a post and hold on as we stand on the top floor, so great is the rolling motion in which we seem to be enveloped. This was so at Saxtead Mill, now disused, but preserved through the strenuous and prolonged efforts of Mr S. C. Sullivan, who is to be congratulated on conducting more visitors over his mill than almost any other owner.

STOPPING A MILL

Before we leave, perhaps our miller will bring his mill to a stop for us; and to be in a windmill when it stops is perhaps the next most impressive experience to being there when it starts. When it starts the noise and motion, beginning with a little creak and shudder, quickly swell to a crescendo, gathering like a hurrying storm. When the miller calls a halt he will carefully release his weights to open the shutters. Very gently he applies the brake when the sails are nearly at rest. The thunder dies away; the clatter of the damsel dies down to a mere tick; the mill ceases to roll and vibrate; all becomes silent and calm, except for a final creak as a strained timber somewhere settles in repose. In fact the proverbial pin might now almost be heard to drop, for it does not need a noisy and boisterous wind to drive the mill, and it may be almost inaudible from within when at last the mill comes smoothly to rest.

Rather different is the story with a cloth-sailer. Then the miller will have to apply his brake first, to the sound of much groaning and cracking, with perhaps a burning odour and a little pall of smoke if the wind is strong and a good spread of cloth is on. Cautiously the miller clings to his brake, and perhaps he will screw down his stones more tightly, until finally the sails come to a halt. Then, hurrying downstairs, he runs to take in some cloth before a gust sets the sails off against the brake—a nasty predicament that may cause the miller injury from an unruly sheet of tarpaulin. Or perhaps the cloth will wind itself around the wind-shaft neck if the sails persist in turning, and he must tie down his brake-rope and scotch the brake-wheel in order to climb the sail to struggle with the offending cloth. So the impressive calm and silence of a halted mill will be lost in the anxiety and excitement of bringing his mill to rest.

Soon we may expect this old mill to go the way of so many others. Who knows? We may find on our next visit that the sails are falling to pieces, the fly is broken, rotten boards are peeping through the last perished and peeling coat of paint; and (as a miller once said to me sadly of his disused mill) 'She's badly goe-d!'

THE JOLLY MILLER

Before taking leave of our mill, however, let us look at the miller himself for a moment or two, for he was often a man of outstanding character, if true to the tradition of his trade.[1]

He formerly had a language of his own, when he would flag to his brother millers—'sails on the cross' if all was well when the mill was stopped for the night, or 'upright' if assistance was required. A miller normally left them on the cross to equalize the strain on the stocks; but some millers momentarily set their sails upright for luck, before striking up for the day's work. And well they might, in hard frosty weather, when they sometimes had to climb 30 ft. up the frozen and slippery sails to wrestle with weighty cloths that would not unfurl, or hammer away at iron couplings solidly iced-up on the patent-sail rods.

Mind you, the miller had his bit of fun. There was Tom Bolton of Twyford Mill in Buckinghamshire who used to let his small mill boy stand on the palm of his hand at arm's length, and could toss the 4-bushel sacks through the first-floor door to save using the sack-hoist—a big man, Tom! He would crack a joke and shout with a laugh which was heard in the village half a mile away, and when the boy laughed he clapped a handful of meal into the lad's mouth. 'What's the trouble?' Tom would innocently ask as the boy spluttered, 'I could do it myself', and then, slipping some meal into his own mouth he would blow it all over the boy!

I was told this by the one-time mill boy, now an old man; so I can claim that (at least) 'I had it straight from the mill'.

Many a miller took a little meal from each sack before delivery (after the days of 'toll' had ceased), and, perhaps to allay suspicion, this was called in the mill 'hanging the cat up'; but some customers got wise to this phrase; a farmer collecting his meal in the miller's absence would slip a little of the *miller*'s meal into his *own* sack, just in case the miller had already done the opposite thing. On meeting

---

[1] In Sussex an employed miller was called the grinder, and the proprietor was the miller.

the miller as he drove off, the farmer would call out 'I've taken up my meal, Joe; *and I've hung up the old cat!*'

At Dale Abbey Mill, Derbyshire, where Miller Smedley and his father and grandfather before him were interested in music, the local brass band used to practise in the windmill itself—not the round-house—while it was grinding.

Often enough when the wind blew in autumn, the miller would work from Sunday midnight to Tuesday evening, Wednesday morning to Thursday night, and Friday morning to Saturday midnight, taking only a few snatches of sleep; and a good windmiller always woke up in bed when the wind rose, getting up in the middle of the night to set the mill going, because the *wind* was his taskmaster and must be taken advantage of whenever it blew. Many a village has at times gone short of wheaten bread because the local mill was becalmed in a waterless district before the invention of the steam engine; and barley-meal bread or even potato bread had to suffice in the crisis of a windless autumn.

Millers of old were adaptable, taking everything in their stride, often helping to build or repair their own mills, coping with heavy weather in a truculent old windmill, dressing their own stones if need be; testing the quality of meal in palm of hand (with the pro-verbial 'miller's thumb'); struggling through snow blizzards to deliver their produce. Sometimes they lost their lives if the winds took control of a post-mill and blew it down, or if a sail struck them as they emerged from the mill door. Sad to relate, one fatal accident has occurred since the last war, in a working windmill. The miller of Terling smock-mill, after working in the mill all his life, was caught up by the machinery while leaning across the great spur-wheel to oil a bearing. (Safety requirements are now to some extent covered by law—for example, all exposed machinery is supposed to be suitably guarded; and although it is hardly practicable to arrange quite fool-proof guards around every piece of mechanism in a windmill, the simplest guard can be an asset if not interfered with.)

Of the personal qualities of the old traditional windmillers and millwrights, one of their descendants, himself apprenticed to the trade, once wrote to me: 'Many of these millers were men of out-standing personality—and invariably God-fearing men too; I wish you would emphasize it—and after years of travel and association with men of all grades of position and intellect I say in my opinion they are a type of character unto themselves, and sterling at that.' I can certainly endorse this view of the average miller—he is sharp and alert always, never loutish and unintelligent like some lesser country-men; in fact his life in some degree depends upon having his wits always about him in an old windmill.

Occasionally a miller has confessed to a sort of fear that his wind-mill would one day kill him. One can understand a lonely miller in a hill-top mill becoming a prey to such feelings as he drives his mill on a winter's night by the light of a hurricane lamp, or with a tallow or rush-light candle in earlier days, but it is less easy to sympathize with the miller who fears to let his sails go, and always ambles along —such a man has got into the wrong profession. (An old man who often worked all night in Brill Mill assured me that lanterns were never used there, although the mill worked until 1916—only tallow candles were used, each of which would last eight hours.)

Tales are sometimes told of millers 'going round on the sails' while they were running, but I have found that even windmill enthusiasts sometimes disbelieve these stories. I therefore wish to state cate-gorically that I have met several millers and former mill boys who went round on the sails in days gone by. They thought little of it because they often climbed the sails to do repairs and make adjust-ments.

THE RESOURCEFUL MILLWRIGHT

No true millwright will ask any man to do a job on a windmill that he would not do himself; and I remember that a favourite phrase of one old millwright when he encountered a sticky problem was 'No windmill ever beat me', and he would cryptically add 'I've done it before I start', meaning that he had it all worked out in his mind. (So far as I know, he never made plans or notes on paper.) 'Show me a job another man can't do on a windmill and I'll do it', he would boast.

Hats off, then, to the little miller-cum-millwright—Robert Young of Kingsey, who moved the Buckinghamshire post-mills; 'Si' Nunn of Wenhaston; John Russell of Union Mills, Cranbrook, miller, engineer and millwright; the late Mr George Foster of North Leverton Mill, Nottingham; and many other handy men of the countryside. They kept things moving in the country villages, variously tackling the job of wheelwright, village blacksmith, iron-founder, bricklayer, carpenter and joiner, mill-builder, perhaps even stone-dresser too. Almost single-handed such men drove an ancient post-mill on the village green, and in wintry gales, hurried out at dead of night to tend a neighbour's storm-damaged mill; whilst on calm days they plied the trade of journeyman millwright up and down the country.

Chapter IX

NEW SAILS AT BRILL
AND OTHER POST-MILL REPAIRS

❧

I CANNOT bring this book to a close without saying something more about a very important side today—windmill repairs. Up and down the country there remains what might be called a fragmentary population of windmills. Most are dead or decaying; some twenty or so only are left to earn money in the traditional way—grinding corn by power of the wind—but many are coupled to a steam-engine or oil-engine. The time may come when all windmills have become at the best museum pieces, and the only work called for on them will be repairs sufficient to keep them standing. It is small consolation that we may live to see larger but less picturesque windmills with 150 ft. steel sails generating electricity to feed into the national 'grid' system.

I have either carried out windmill repairs from time to time or have been interested in what has been done by others in the matter, and I have introduced the word 'millwrighting' into the title of a windmill book for the first time, because I proposed to describe both the construction of windmills, and also some of the reconstruction work done.

Let me therefore tell something of the work done in repairing and reconstituting the windmill at Brill; and I shall use, if I may, the technical terms of the old millwrights with a minimum of explanation, because even in the language itself something is contained of the labours involved.

Brill windmill had been presented to the Buckinghamshire County Council by Sir Henry Aubrey Fletcher; and early in 1948, I telephoned my friend Mr Frank Boothman of the County Planning Department and learned that, as elsewhere, difficulty was being experienced in finding a millwright who could come to Brill, or a builder who was not fully occupied with housing.

The first essential was to replace some missing boarding on the roof, and I proposed that we should offer to do this ourselves. This Mr A. H. Prince, A.M.Inst.C.E., County Planning Officer, agreed to; and at his suggestion I submitted a report on all the windmills that might be worthy of preservation in Buckinghamshire.

STORM DAMAGE

A visit to the windmill revealed that in the storms of March and Easter 1947, in which a sail was broken off, the turning of the remaining three sails had split the brake-wheel, which was chained to the spindle-beam. The chain had torn this spindle-beam out of its socket in the wall, also causing the wall to be wrenched outwards, as a result of which the near-side end of the weather-beam had almost come adrift. We decided that three tie-rods were necessary—one across the mill, close to the spindle-beam, another from the weather-beam back to the rear corner-post, which is 5 ft. short of the tail of the mill, and a third from the crown-tree down to the bottom floor-rail, since one lower side wall was sagging and threatening to part company with the side-girt (but this floor is now supported from the ground). In the meantime we took the roof repairs in hand.

PRELIMINARY WORK

On 6 February, being in possession of the tie-rods from Messrs Roblin's machine-shop in Aylesbury, we drove to Brill in a 50-mile-an-hour gale, to find the horizontal pair of sails swinging to and fro, the wedges having disappeared; so that the sails rocked the mill back and forth.

I decided to install a rod behind the spindle-beam immediately, which we did with some trepidation in view of the alarming motion of the mill in the storm; and I delayed our return to the shelter of the village another few minutes in order to climb through the brake-wheel and knock some oddments of wood into the poll-end to steady the sails sufficiently to lessen the immediate strain on the mill.

All the weatherboarding we fixed later from inside, the last board in each gap being drawn tightly into place with copper wire, since we could not reach to nail it. Later, it became obvious that the remaining sail-stock was weak, and the straps holding the sails were bent and rusted away; so I advised Mr Prince that the sails should be removed for examination before another one came adrift and did more damage.

REMOVING THE OLD SAILS

A worm-gear and sling chains and ropes were hired, and Mr Clarke later sent me his light rope-tackle. The worm-gear, owing to its inaccessibility, was less serviceable than a rope-tackle, but we also had a single block pulley and some more rope, and after removing the odd sail we pulled the remaining pair round, and roped the lower tip of this stock to the cross-trees to prevent the top sail coming down when the lower one was removed, as the brake was displaced. (This was restored in due course, but at the moment we were anxious to concentrate on the sails.) As this sail had been up 60 or 70 years without being taken down when repaired, all the nuts and bolts were rusted up solid, and we had to saw through the ¾ in. bolts. In order to lift the sail to ease the last bolt out, a car jack was stood on some bricks, under the sail-tip, and some slip-knot ropes holding the whip loosely to the stock were then pulled, and the sail lowered and removed to the late George Green's yard, across the common, from which eventually we organized all the work. Now we let the top sail down by paying out our main rope, but finding the bolts rusted as before, I decided to release this last sail and the stock together, and hitched both the big rope-tackle and our single pulley and rope to the stock, suspending them from the outer box of the canister. The sail and stock slipped down to the ground easily, and we levered the sail tip to one side with a pinch-bar, with two men to each rope, but these were not around the cross-tree because of the strain this would involve; consequently when the stock (only 30 ft.) released itself from the poll-end, it lifted the four men off the ground, and sail and stock were at our feet without delay (Pl. 29a). No harm was done, as I had all my men out of range of the sail and stock; it is a good rule in millwrighting never to have anyone underneath something which might fall on him if a rope or chain snapped. The sail was unbolted on the ground, where we could apply more force to it.

INTERNAL REPAIRS

Then came the internal renovations. I refixed the broken 'windows' (wooden) and the sack-traps, repaired the front corner of the mill, fixing an iron coupling from the top side-rail to the weather-beam, and fitted up the auxiliary gear take-off. I repaired the step ladders, blacked all new iron-work, and attempted (with some assistance) to close the split in the brake-wheel; but we could not draw up this split, as it seemed to be jammed with splinters, and would need opening up and cleaning out. I reinstated the governors, cemented beneath the horns of the main-post, as it had already sunk on to the cross-trees, filled the holes in the cross-trees with cement to keep out the wet; and I put an oak wedge between the cross-trees so that the weight of the mill is now communicated directly to the ground.

MAKING NEW SAILS

We now made a momentous decision: in fact, to fit a slightly modified full-sized set of sails and stocks, offering less resistance to the wind at the tips than the old ones. I prepared the new design on Mr Boothman's plan of the old sails. Seventeen sail-bars spaced at 18 in. were provided, instead of eighteen bars 17 in. apart, for the greater convenience of the timber merchant when measuring, and the uplongs were omitted from the last few bays, the outer end of the wind-board dispensed with, and the hemlath drawn in at the tip to shorten the sail-bars. The sails are 54 ft. span and 6 ft. broad, as before, and I provided the timber man with a 2 ft. model of one sail to illustrate the principle of obtaining the warp, or angle of weather, through the offset of the mortises in the shaft; I also provided a graph giving the angle and position of each bar, and a drawing of back and front of the whip, showing how to set out the mortises on each face. The tips of the sails were brought forward a trifle (as mentioned in Chapter 3), in accordance with Mr Jesse Wightman's theories (with which I agree), and the hemlath was a fairly evenly graduated curve as seen in Lincolnshire. This does not coincide with the ideas of some millwrights; however, the new sails actually made several revolutions with no cloths, and the brake hard on, in a 70 m.p.h. gale on 9 February 1949, when the brake-wheel had not been chained, although, so far as is known, they did not move in an 80 m.p.h. wind the previous month (when gusts of 93 m.p.h. were registered in Wales and of 65 m.p.h. in London). On both these occasions the wind was from the north-west—not a tail-wind—and the sails ran *with the brake*, not backwards, which can more easily happen.

Messrs Reynolds of Waterside, Chesham, cut the whips with a band-saw on a travelling bench, and mortised them by machine, and so well was this work carried out that not one sail-bar had to be trimmed finally to fit, and the sails when completed had the intended curvature or 'weather'.

One of the stocks, cut elsewhere, was made ½ in. too thick and had to be trimmed down. On the advice of Mr Walter Rose of Haddenham, and Mr Prosser (the builder at Brill who very kindly rendered assistance and provided some materials free of charge), I recommended larch fir for the stocks, in the absence of pitch-pine, but the County Council were unable to find big enough larch trees on their estate, and so had to use a shorter-grained and more brittle fir.

Mr Jack Blake, a Council man at Brill, assisted with the assembly of the first sail (Pl. 29*b*). All sail bars were nailed, and the uplongs, hemlaths and back-stays were all similarly nailed and clenched at the back. Mr Blake did the other three sails to the pattern of the first. The bolt holes in sail-stocks and whips we made by measurement, as we could not easily shift the sails and stocks in Mr Green's yard to lay them together for checking. We then noticed that a ½ in. slice would have to be removed from the first 16 ft. of one of the stocks in order to pass it through the canister. Cross-cuts were therefore made at 6 in. intervals along the stock and the timber sliced off with a broad chisel by Mr Blake, who then creosoted all the sails and the stocks (I have since learned that a man experienced and skilled with the adze might have trimmed the stock even more adroitly.)

FINAL PREPARATIONS

We decided that the cross-tree and quarter-bar beneath the head of the mill must be strengthened to receive the weight of the new sails and stocks, which was estimated at 3 tons, although Mr Reynolds said they would lose nearly one-third of their weight in 12 months owing to evaporation of the sap. (The old sails were probably well under 2 tons when they went to pieces, as they were perished.) A 14 in. thick slab was accordingly fitted beneath the cross-tree, and a 9 in. one under the quarter-bar, with a double strap or link supplementing the existing one (which is secured by a bolt and a gib cotter), at the foot of the quarter-bar, to hold all these timbers together, and to prevent any forward movement of them, as the new bottom slab is behind the brick pier.

Some people imagined that the new sails would serve only to ornament Mr Green's yard indefinitely; the idea of getting them up on the mill appeared fantastic to them; though considerably more than a century ago, it should be remembered, the 120 ft. tower-mill at Gorleston was fitted with 84 ft. patent sails, weighing 5 or 6 tons.

So I brought Mr Blake and his team, John Boughton and Fred Collett together on 30 September 1948, and we suspended the rope-tackle from the canister, with the single pulley alongside (used in conjunction with each other to multiply the leverage), together with the worm-gear in case of need.

Transport of the 30 ft. stock across the common presented something of a problem; but Mr Prosser had lent me a hand-trolley consisting mainly of a pair of iron wheels and axle from a threshing or similar box, with a stout handle affixed, and we levered a stock up some sloping poles on to this, wired it on, and trundled it over the common, lifting it directly off the trolley with our tackle (Pl. 29c).

When the canister began to take the weight of the stock it was apparent that the main-post was so weakened by worm or dry rot that the mill-body would also need support, and we had to cut two 11 ft. posts, 13 × 6 in. section, and erect them beneath the two shuttle-trees, which made a wonderful difference. It is said by those who have since stood beneath the mill in an 80 m.p.h. gale that she is now firm as a rock.

ERECTING THE STOCKS

Next morning the first stock was successfully erected. Our tackle was coupled to the front face of the stock, with a sling chain about 6 ft. down from the tip, and we levered the stock into the inner eye of the canister with a crowbar passed down through the eye; and the recently applied creosote caused the timber to slide very freely against the iron instead of binding.

Four or five men—willing helpers soon gather round when an unusual job is being done—then hauled up the stock clear of the ground with the rope-tackle; and we attached another sling chain and coupled up the worm-gear, which was worked from a ladder because of the absence of trees to which to hitch the rope-tackle. We soon had the stock home up to the shoulder, and with a rope already attached to the tip, we pulled it round to the horizontal position in order to insert the poll-wedges.

On 2 October, a Saturday, the Council men were not available, so I engaged two other assistants, knowing that extra hands would surely be applied to the ropes when necessary if one or two helpers were available to get *something* moving.

We fixed our rope-tackle on the new stock at one side of the poll-end, with the pulley-block alongside it as before, and the chain-gear the other side. The stock failed to enter the eye, and heeled over towards the mill too much, so another single block was suspended over the nose of the poll-end and a line passed over it from the stock; someone on the ground could then pull the tip forwards while the foot was levered inwards with a pinch-bar.

HOISTING THE SAILS

The first sail was brought up and laid with its heel pointing towards the mill. A sling rope was used instead of a chain, to save damaging the sail-bars, and two men hung on to the outer end of the sail to make the heel rise up sharply, when the sail was lifted by the rope-tackle, so that it would not drift in towards the mill-body and lodge behind the sail-stock. Having coaxed the sail in front of the stock pretty easily, with the two men still guiding the tip of the sail (Pl. 29 *d*), it was only the work of minutes to slide it up the stock to the desired position. A helper then climbed the sail (which, on such a low-built mill as this was readily kept steady by holding the tip), and inserted the bolts and fitted the two straps after a little chiselling at the corners of the stock. Erection of a 'common' sail-frame is a very straight-forward job compared to getting a stock in place; but we now had to get it up to its topmost position, in order to attach the opposite sail of the pair; and we presently brought the stock partly round with some preliminary pulling at the ropes attached to the tips of the stocks. Mr Les. Worthington, the artist and engraver, who had brought his motor caravan specially from Liverpool to see the work done, then very kindly attached the rope to his motor and pulled very slowly and smoothly till I signalled to him that the sail was nearly vertical.

The mill-brake was applied and the second sail affixed like the first; it was of course possible to unhitch the motor as soon as this sail was clear of the ground, with its weight on the stock.

The shaft of the third sail had warped 3 in. out of the straight, but it easily pulled up to the stock when the nuts and bolts were tightened. A Council lorry had now brought some materials, and with several helpers we might have got this sail to the top position, but we decided to hitch the rope to the lorry (Mr Worthington having had to leave) and the fourth sail was attached, the same as the second.

OFFICIAL APPROVAL

Just then, on the afternoon of 4 October 1948, the County Planning Officer and other officials arrived, and were not unnaturally pleased to see the results of our handiwork. The district foreman was particularly surprised and impressed, for he had previously called when we were having to shore up the mill, and had gone away thinking we would never get the sails up. Later, it is understood, the repair of Brill windmill was quoted by a lecturer to an architectural class at the Polytechnic as an example of a well-executed preservation job.

Our poll-wedges, which Mr Prosser kindly cut and gave to us free of charge, were of oak—some millwrights favour hornbeam or apple wood, the same as for mill-cogs, but Mr Prosser considers that the quality and texture of the wood is the main consideration, and oak is more durable for exposure to the wet, and is in fact used in a watermill for cogs which come in contact with the stream.

The backstays of the sails were fixed later. Eventually Mr Boothman and I creosoted the roof of the mill with a stirrup pump, applying 10 or 12 gallons. We then learned that Mr Blake was injured through a ladder slipping and falling down, the day after we finished the sails; and that while still on sick leave, he had constructed an excellent model of the windmill.

GENERAL MAINTENANCE

Shortly afterwards Mr Harry Wootton, a helpful local volunteer, helped me to fix stranded wires between the tips of the sail-stocks (to equalize the strain), and also a central bowsprit in front of the sails, with strands of fencing wire to each sail to steady them against head winds. These wires were tautened by binding them to the sails with lengths of motor-cycle brake wire secured with staples, and inserting a bar between the strands to twist them. It was essential to have these strainer wires at Brill, because of the brittle nature of the timber provided, although some persons in the village were opposed to any deviation from the original details of the windmill.

Later I received Mr Blake's message that the sails had revolved in the storm of 9 February, and I visited the mill to check the brake and chain the wheel. So great was the wind, they told me, that some passers-by were busy keeping their own balance, and did not see the sails revolving.

In due course I painted several explanatory notices within the mill, tightened the poll-wedges and the wires, creosoted the interiors of the stone-cases (which were very badly worm-eaten), and lifted the marginal floor and spindle-beam with a car jack and fixed them. Brill windmill—better built than the midland post-mills though not aspiring to Suffolk standards—has the side-girts secured with two upright bolts, the waists of which are made to hold short dog-irons which emerge from the side-girt to lay atop of the crown-tree, where they are in turn spiked down with big forged nails. One of these spikes had rusted and broken, and so I drove in another, making it serve at the same time to anchor a tie-rod coupled to the head of the mill on the far side (where the corner-post was cracked) for additional strength.

RENEWING THE ROUND-HOUSE AND COMPLETING THE JOB

We now decided to construct a new round-house; and I measured the foundations of the old one (21 ft. 6 in. diameter) and fixed the position of further supports for the body. Two days were also spent creosoting the sails, doing the inner portions from the sail-stocks with a tar-brush, and the outer ends with a 3 in. paint brush, by climbing each sail.

A search was then undertaken for bricks and workmen, and we approached a retired bricklayer, Mr Arthur White, whose father built the original round-house for the late Andrew Nixey about 1865. Mr White estimated that 5000 bricks—an eight-wheeled lorry load—would be wanted, and he agreed to do the work if someone would fetch and mix the cement. The cement mixing was entrusted to two men because of the amount of fetching and carrying involved. We were lucky to find the necessary bricks in the Brill timber yard, the site of Brill's last of many brick and tile yards. The final big brickyard eventually put the 'little men' out of business; but the great brick walls of its own kiln are now being demolished, and Mr Wiltshire, the present proprietor, gave the bricks free. He also provided the timber for the round-house roof.

Meanwhile, half a dozen 6 × 6 in. oak posts were sent to the mill, and I erected and wedged these on concrete bases, so that the mill cannot roll or strain the main-post unless the tail-pole is allowed to break away from its moorings, in which case the mill *might* suffer a twisting strain from a rough gale.

The old round-house foundations of local stone were exposed and levelled up by Mr White, who then built the walls; and a carpenter was found in the village to construct the new round-house roof framing, to which cedar-wood shingles were fitted by the makers. I recommended that the cross-trees and quarter-bars should be creosoted with a stirrup pump meanwhile (which is preferable to using this pungent liquid in a closed and dark space where thorough dressing in the corners and crevices could not be ensured), but the order was not given for this to be done. The round-house was completed with an excellent oak door salvaged by Mr Green from the oldest house in Boarstall village, and the ladder is now fixed inside, with a trap in the mill floor. Thus is the mill restored to its general appearance during the last 50 years of service, when it had a round-house, but the permanent ladder (undesirable because of children climbing on it) and the cart-wheel on the tail-pole are missing.

I made and fixed a date panel 1668 (the original date of building) to interest visitors; although a document has since come to light— an account book kept by one William Snell in the eighteenth and nineteenth centuries—containing the following rather exciting but cryptic entry, for no further reference to the matter appears:

1757

Nov. 2—Expenses of Rebuilding a Windmill which was
blown down at Brill, exclusive of value of old
Materials, and which Mill is now Lett to John
Atkins at per Ann. £9 . . . . £175 5 0

The same book reveals that William Adkins was the occupier of the mill in 1805; while the *Posse Comitatus* recorded the two millers of Brill as William Adkins and James Parsons. This identifies the above mill as the existing one, since the other (demolished 1906) is well known to have been Parson's Mill; and it is certainly clear after examination that the present mill was reconstructed while retaining a seventeenth-century wind-shaft and main-post.

OUTWOOD MILL, SURREY

A major repair job was undertaken by Messrs Hole and Sons, the millwrights of Burgess Hill, in August 1952, at Outwood, Surrey; and the author's pleasure at finding this work in progress, on a chance visit, may easily be imagined.

The firm were then renewing the weather-beam, neck-wear and prick-post with some very nice pieces of oak, and Mr Jupp, whose family have worked the mill for 200 years, very kindly permitted the work to be closely inspected by Mr Denis Sanders and myself.

Outwood Mill, the fourth oldest in England (pre-dated only by Pitstone Green Mill, Buckinghamshire, 1627, Bourne Mill, Cambridgeshire, 1636, and Lacey Green Mill, 1650) is by far the oldest still at work, having been erected in 1665; and it is recorded that the builders of the mill watched the Fire of London from the roof as they finished their task the following year.

A tall smock-mill stood alongside, but she was little more than 100 years old and was derelict, never having been a very useful mill. The post-mill has a single-floor round-house, and is based on a 'head-and-tail' layout, with an exceptionally large wooden wind-shaft, 22 in. diameter at the clasp-arm head-wheel, and 20 in. octagon where the iron-centred, wooden tail-wheel is attached.

As the repairs were carried out with the four sails in position, it was necessary to shore up the windshaft and sails very securely whilst the weather-beam and prick-post, which normally support them, were entirely removed. To do this, the sails were supported by two newly cut 12 in. thick fir trees lashed together under the canister-head, and standing on boards in shallow holes in the ground, to form a pair of sheer-legs. The sails were prevented from turning by means of the brake, and by having long guy ropes attached to their tips.

143

To preclude any forward movement of the wind-shaft in a tail-wind two ½ in. stranded wire cables, 50 yards long, were tightly bound round the wind-shaft behind the brake-wheel and passed through the side windows, and one from the tail of the shaft out of the back window. One wire was fixed to some trees, one to some iron-work behind a shed, and one to a staple and bedplate (buried in the ground) as used for securing telegraph poles; and they were moderately tensioned with turn-buckles but not enough to strain the mill.

The new weather-beam is bolted and secured with heavy angle-irons to the top side-rails and front corner-posts. A new neck-beam is bolted down; and the new prick-post, of $9 \times 6\frac{1}{2}$ in. oak, fitted after the weather-beam was installed, is mortised up into the weather-beam and attached at the foot to a heavy iron angle-plate bolted on to a new centre-piece on the bottom front rail. The angle-plate is also held up by two tie-rods from midway up the front corner-posts.

WETHERINGSETT MILL

One of the most complete reconstructions undertaken within living memory, amounting practically to the erection of a new post-mill with an assortment of parts from its predecessor, was the building of Wetheringsett Mill, mid-Suffolk, in 1883, which incorporates the last new main-post ever made in England. The old mill had been blown down on Black Tuesday (18 January 1881) when the greatest wind within living memory, accompanied by a great snow blizzard, wrought so much havoc in East Anglia and Kent and Sussex. Farther inland the snow is remembered more than the wind, but the wind blew with destructive violence all around the coast and 50 miles inland.

The late Mr E. Aldred (cousin of the better-known Steve Aldred of Saxtead Mill; Pl. 26a) and his father, Frederick Aldred, had Wetheringsett Mill for some years prior to 1881, and the present mill has always remained in the family. Father and son helped to build it, although Mr Sam Clarke and Messrs Whitmore and Binyon were all involved. It appears that Whitmore and Binyon undertook the job

and drew up the specification, and Sam Clarke was engaged to pre-
pare and erect a new main-post and crown-tree, etc. The main-post
was turned from an oak tree selected from the hedgerow of the
Ashfield Swan public-house meadow, grown on strong clayey and
stony ground, and the crown-tree was from another oak a quarter of
a mile away. The main-post is now rotten at the top owing to
'worm, but otherwise the whole structure is still pretty sound,
although a crack has developed in the round-house, and extends in to
one of the piers.

PITSTONE GREEN MILL, IVINGHOE

In 1954, whilst inspecting this windmill, I made an important and
interesting discovery—the carved date on a wall timber, alleged to
be 1687, had been misread, and is in fact 1627, *making this the oldest
dated windmill in England.* Mr O. Evans of Tring Flour Mills, and
Mr J. D. Hawkins, whose father formerly owned the mill, both
agreed with my reading of the date after a most careful scrutiny, and
also confirmed that this timber is contemporary with all the other
framing timbers in the mill (Pl. 21c).

The outcome of our general inspection was that we discovered
the mill to be in urgent need of shoring up, mainly because one of the
quarter-bars is sinking into the decayed end of a cross-tree, as a result
of which the main-post is out of perpendicular. This was brought to
the notice of the owners—the National Trust—who I understand
were advised by the S.P.A.B. to approach me to supervise some
repairs; and we thereupon met the Trust's local agent, Mr R. P.
Coles, at the windmill.

To remedy the trouble temporarily, we have inserted two 16 ft.
pitch-pine baulks, which are tenoned into the 'dummy' mortises in
the main-post (used for supporting the post during erection) with
the feet of the new timbers wedged against wooden plates at the
base of the round-house wall. As soon as possible we plan to insert
permanent oak posts on the lines of those at Brill; and the question
of making and fitting new sails will then be considered.

HEIGHTENED POST-MILLS

It was not unusual in the early nineteenth century for the Suffolk millers to have their post-mills raised bodily on to much taller piers, partly to catch the wind better and partly, in a competitive spirit, to possess the tallest windmill in the district.

This was achieved by jacking up the mill bit by bit, piling timber baulks beneath, and building up the brick piers as the work proceeded, so that in two instances—Honington and Thorndon—the roof ridge is believed to have attained a height of 55 ft. from the ground. In more than one case, however, a misjudgement occurred, and the windmill, overbalancing, crashed to the ground and was wrecked.

RENEWAL OF SAIL-STOCKS

Preparation of new sail-stocks in olden days presented more of a problem than it has done recently, when large 'travelling' saw-benches are to be had; and within the last fifty or sixty years a new pair of stocks was cut to shape on the site, from squared pitch-pine baulks, at Haughley, near Stowmarket. The sawing down lengthwise of these baulks, by hand, to obtain the correct taper, was considered to be a big job even amongst the old inhabitants, used as they were to heavy sawing. No rigid bench was at the time available, only trestles; and some millwrights would almost certainly have cut the surfaces and trimmed them down with an adze instead of sawing.

It has, of course, become a common practice during the last few decades for sail-stocks (and sails) to be transferred from dismantled mills to working ones; probably half the windmills now standing have second-hand sails and stocks, and it is today almost impossible to find a disused mill with stocks good enough for service elsewhere.

Chapter X

WINDMILL REMOVALS

❧

MANY are the stories of the removals of windmills in Sussex and East Anglia—of hauling them bodily along with teams of oxen or horses; of being stuck in the mud overnight; of upsetting the mill altogether as it bumped and rocked along some ill-formed cartway; but in the Midlands (at any rate) the actual mode of removal is almost completely unrecorded.

MIDLAND POST-MILL REMOVALS

Many post-mills were moved in the Shires; this is well known; but almost certainly they were not moved bodily. None of the oldest millers interviewed in the south Midlands before the war had ever heard of this being done, and the general opinion is that the steep hills upon which many windmills originally stood, and the very bad or non-existent roads, rendered such a task impossible. Presumably these windmills were carefully dismantled piece by piece, since this is the only convenient alternative to removing the sails and machinery, and lowering the whole carcass with the aid of piles and jacks on to a specially made truck. Identification numerals were not (so far as is known) found on the parts of the re-erected post-mills of the Midlands (as in certain smock-mills); but this might be because all the main timbers are easily recognizable, no two walls in a post-mill being alike; and the smaller studs were no doubt renewed, as some of them were at Thorpeness Mill in 1923–4. Had they been bodily moved, with or without main-post and cross-trees, the older millers would have known—just as many millers elsewhere knew the number of oxen or horses employed, usually from 16 to 24, the time occupied, and the various misfortunes and adventures encountered *en route*; and some particulars of the specially improvised truck or trolley used would surely have been forthcoming.

BODILY REMOVAL OF POST-MILLS

If the main-post and cross-trees were to be retained, they might be removed complete with the mill carcass; the cross-trees would be shored up, brick piers demolished, and the trolley run underneath the structure, which was let down bit by bit with jacks and levers. Sometimes in Suffolk two 'drogues' (or timber wagons) were lashed together side by side, two arms of the cross-trees being rested on one, and two on the other.

When removing the carcass only, and renewing cross-trees and main-post, the body would be shored up, main timbers let down and removed and the body placed perhaps upon a single drogue; and in later years a steam traction engine unromantically handled the load. In Sussex, 150 years ago, massive, many-wheeled trolleys like the modern tank-carriers were specially built for hauling windmills over the Downs, where ordinary vehicles would probably have sunk in and become bogged.

One of the best known examples is that of a Brighton post-mill, moved bodily from West Street to Dyke Road in 1797, for Mr Streeter, the miller. On the engraving that has been several times reproduced, it is stated that eighty-six oxen were employed to haul the mill, and they are shown harnessed six abreast—oxen were preferable to horses, especially on rough ground, because they take up the drive more steadily. It is said that Mr Streeter's mill was moved 'whole and literally as he worked her', which presumably meant that the millstones and gear (usually removed) were left intact; but the engraving, at any rate, shows that the sails were off.

Usually windmill removals began at dawn on a summer's day, to avoid traffic; but the wagon might be ditched, or bogged in a ford; the mill might even topple over, or a stumbling horse would be trampled on and killed in rounding a bend—an ever-present risk with twenty or more horses. Trees had to be cut down; and delays sometimes amounted to a day or a week; at other times, however, a removal was accomplished in a matter of hours.

REMOVAL OF ST JAMES'S MILL

Messrs Adams and Ball of Huntingfield moved a mill 'buck' from a site between Harleston and Starston, Norfolk, to St James South Elmham, Suffolk; Adams being an engineer and Ball a millwright. They seem to have been in business separately, but were doing some good work together at that period—about 1870; and Mr Bullock, who was working for Ball at that time, gave me the following particulars:

A miller named Clarkson commissioned the work, and Mr Bullock was at St James making preparations, when the buck arrived, minus its top, on a drogue to which a sort of frame had been fixed to afford more support; and he helped to erect the mill with new cross-trees and quarter-bars. Adams, he said, wanted to convey the buck on its side, on two drogues lashed together; but Ball insisted on having it upright, which was the usual way; and when they reached a low railway bridge Adams again proposed laying the buck down, which would have meant practically starting all over again, but Ball would not have it; he removed the top, which, of course, was refitted later. The new cross-trees, with quarter-bars and main-post, had first been erected to ensure that the mortises were accurate, and were taken down again to make way for the buck. The buck was levered up with eight jacks and numerous pinch-bars; and they avoided the disaster of having it topple over.

Then the main-post was inserted from beneath, and the cross-trees were again set up. With quarter-bars inserted in the cross-tree mortises, the post was set down on to them and then the buck lowered till the crown-tree socket settled on the pintle, firmly jamming home the quarter-bar tenons. This safely accomplished it remained only to replace the roof, hoist the sails and fly, etc., and install the internal equipment, the same as if completing a mill newly built.

THE THORPENESS REMOVAL

Only once in present living memory has a post-mill been removed and re-erected *as a working mill*, although at least one other has been reconstructed on a new site (Madingley, Cambridgeshire) for ornamental purposes.

The proprietors of a very extensive estate near Aldeburgh, Suffolk, decided to have a real old-fashioned post-mill for pumping water alongside the well-known boating lake at the model village of Thorpeness. They happened to have several windmills on the estate, both for grinding corn—Westleton and Aldringham post-mills—and for draining the marshes—East Bridge and Minsmere 'Black Sail' smock-mills.

They decided upon Aldringham Mill, being, it is believed, not altogether satisfied with their tenant, though he loved his mill and was a very keen miller. Mr Amos Clarke of Ipswich was sent to inform the poor man that he had come to dismantle his windmill *at once*; there was great consternation; but the estate was adamant, and the task was begun then and there.

The job had to be completed, come what might, in wintertime, as no renovations and improvements were permitted to be seen in progress by the summer visitors to Thorpeness; and the elements did not fail to make themselves felt, for Mr Clarke had to paint a good part of the re-erected mill in his overcoat, while the snow lay all around.

Mr Ted Friend, who is over 80 at the time of writing, was the estate carpenter and millwright for many years, having been apprenticed to 'Si' Nunn of Wenhaston towards the close of last century; and he was deputed to arrange and supervise the job whilst the Clarkes—father and son—were engaged to do the dismantling and re-erecting. They removed all the straps, bolts, dog-irons and so on, as they came to them, and drilled out all the dowel-pins. Mr Friend decided that the new pump must be driven by a shaft passing down through the main-post from the sails (which is, indeed, the only convenient arrangement), and he had to make a model to convince the proprietors that the shaft must be bored.

The site was not very handy for the millwrights in wintertime, being on open sand dunes on the east coast; and the lack of trees or posts led to a 450-yard rope being made of manilla (they would have preferred hemp). Mr Clarke still has half of this rope in new condition, having sold the other piece—and a finer rope has probably seldom been seen in a mill yard. For drilling the 18 ft. main-post a special long bull-nosed auger was purchased (Pl. 20c); it will cut straighter than a screw auger (which is easily turned askew by the grain of the wood) and is faster to operate than the old-fashioned shell auger, although this no doubt would have done the job.

Mr Friend used the old upright shaft with the wallower and great spur-wheel; and from the wallower a short layshaft was mounted lengthwise of the mill, with a *disk* (not a crank) on the end of the shaft, beyond the tail bearing. A connecting rod was coupled to a crank-pin fitted into the disk, with its little end attached to the driving rod passing down through the main-post. The spur-wheel served only as a flywheel.

Drilling the Main-post

Mr Clarke proposed to convey the complete buck upon the sail-stocks laid lengthwise on a timber wagon; but the whole mill had finally to be dismantled piece by piece and reassembled, because the estate banned its bodily removal, thinking this would cause damage to the unmade roads.

There was in any event the main-post to drill. The drilling job was started with a shell-bit (something like a small shell-auger). The 2 in. bull-nosed auger, with 9 in. of screw and an 18 ft. shaft, was then set up on benches, and the job was worked half from each end, to minimize errors and reduce the risk of breaking the auger. Boring occupied the Clarkes two long and busy days. At the centre of the post, the two bores were a quarter of an inch out of alignment, and a quantity of parings to some extent blocked the hole. Someone looking on hit upon the bright idea of fetching his air-gun, which he fired through the hole, effectively clearing it! Two long rods were then coupled together with a cutter fixed with a set-screw at the joint, and the rods were worked backward and forward till the irregularity in the alignment of the hole was straightened out, making three days' work in all.

Reconstruction

Now the re-erection was proceeded with in the usual way, much as if a new mill were being built, a few of the studs and other small timbers being renewed, and also the outer boarding. Everything went well until it came to the pumping gears, but owing to the time limit on the job, some existing gears out of another mill were installed, and on starting the mill it was found that the pump would not work satisfactorily at the rate the sails would usually drive it. Then there was a scamper to make a 'pattern' in time for a new crown-wheel to be cast and installed.

This seemed to solve the problem until one day the top of the driving rod, protruding above the main-post and crown-tree, and coupled to the connecting rod, buckled over, because no guide-bars or slides had been provided to direct the rod in its downward thrust. This certainly was a predicament, but Mr Friend overcame it by means of a blow-lamp with which he straightened the rod, then he fitted a guide-shaft reaching across from the mill wall to the cross-head, as provided for the three connecting rods in East Bridge Mill. This does not, by the way, secure a truly vertical up-and-down motion at the top of the rod, but it keeps the cross-head moving within a slight arc, the radius of which is equal to the length of the guide-shaft.

SMOCK-MILL REMOVALS

Methods adopted in conveying smock-mills from place to place are, curiously enough, on record in the south midlands, but not to the writer's knowledge in the east or south so far as full-sized grinding mills are concerned.

Lacey Green Mill, Buckinghamshire, transferred from near Chesham in 1821, was dismantled piece by piece and the beams incised with Roman numerals and also with curious crossed strokes and squares, something like stonemason's signs. All the machinery (constructed in 1650) was reassembled and is still intact, and steps were taken in 1936 to preserve the mill, which is of average size and far more massively built than the standard Kentish smock-mills; but it has since become in need of further attention, partly because of wilful damage.

Salt Hill, or Chalvey Mill, Slough, was removed to the Chatham district in 1848, in sections, by the simple expedient of sawing from top to bottom of each cant-post, thus keeping the several individual sides intact until they could be clamped together again at the new site. Here they were packed with tarred felt in the joints, ensuring that the interior was snug and weatherproof when the straps and bolts were tightly drawn; and a similar procedure seems to have been adopted with a small marsh-mill moved from Dilham (Yarmouth) to Wymondham (probably in 1858), where it became a corn-mill. A small Oxfordshire smock-mill, and several similar ones elsewhere, were cleared of machinery and floors, and the carcass laid bodily on its side on a timber wagon for removal.

TOWER-MILL RECONSTRUCTIONS

So far as is known to the writer, no tower-mill has been completely reassembled on a new site; but an example has come to light of a smock-mill being dismantled at North Marston, in the Vale of Aylesbury, and re-erected as a tower-mill. For this purpose, the large local white roughly dressed stones of the ground-floor walls were all numbered with crude tarred figures which were remembered by old inhabitants before the First World War; but the stone was soft, and the markings have now crumbled away. The upper walls, previously of timber, were reconstructed of red brickwork, giving the complete mill a pleasing appearance.

Various instances are recorded of the interior machinery, floors and fittings from a dismantled tower-mill being installed in a new brick shell elsewhere; perhaps the best known is the transfer of the equipment of a mill upon the site of the proposed extension of Boston Docks, which was incorporated in one of the largest and best proportioned eight-sailers in England, at Heckington (Pl. 11). This splendid mill, which was adapted for timber-sawing as well as grinding, was maintained, until the war made things too difficult, by the miller and his men, and was in fact the last working eight-sailer in this country. It is to be preserved.

OLD-TIME CELEBRATIONS

Windmill removals never failed to provide a measure of excitement for the local inhabitants—even in those spacious days, when the sight of a great white post-mill body approaching down some country lane behind a team of twenty or more struggling and panting horses, and accompanied by a crowd of no less excited and panting men and boys, was not by any means unknown.

One such removal, believed to have taken place on 5 November 1853, was the transfer of one of the two tallest post-mills in England from Sapiston to Honington, the mill being 55 ft. high to the roof ridge, as she stood at Honington, the same as Thorndon Mill, farther eastwards. Both were in Suffolk and both are down, but their round-houses, three-storied with shallow roofs, remain, looking like small brick-built gasholders; and the Thorndon round-house even by itself is a striking building, being of a warm red brick and nicely placed alongside a neat little mill-house.

The Honington removal (which may have included the main-post and cross-trees, as twenty-four horses were required, even with millstones removed to reduce weight) was the subject of a country ditty, though only the first two lines seem to have survived until 1939: they ran as follows:

'Here come young George Cutting, all rausel and tear;
Let me hang on to my old black mare....'

Fifty or sixty years ago the rhyme was sung by the children; but memories grow short and now even the old folks seem to have forgotten this last record of a vanished rural scene.

GLOSSARY AND INDEX

PEAK STONE	Derbyshire stone of millstone grit for barley, etc.	13, 44, 68
PECK LOAF	Seventeen and a half pounds	110
PEG MILL	North country name for a post-mill	5
PENTHOUSE	Pointed turret on fan stage of smock- or tower-mill	5
PIERS	Brick or stone piers supporting cross-trees of post-mill	17, 82
PILLOW-BLOCK	Iron block resting on neck-beam and carrying neck-bearing	35
PINCHING SCREWS	Set screws for centering a spindle in its bridging box	48
PIN-GEAR	A trundle wheel (q.v.)	11
PINION	The smaller of a pair of bevel or spur cogs	11, 43, 56
PINTLE	The pivot centering a post-mill on top of main-post	20
PITCH-CIRCLE	Imaginary circle on ends of cogs, cutting them at the main point of contact with cogs of the complementary wheel. (The 'pitch' is the distance between two cogs measured on this line)	91
PIVOTAL PLANE	Plane in which millstone is balanced on spindle	48
POINTING ROPES	Ropes for reefing a 'common' sail	39, 96
POLLARDS OR 'THIRDS'	A lower grade product of flour dressing	107
POLL OR POLL-END	The canister (q.v.) on a wind-shaft	36, 87, 138
POLL-WEDGES	The wedges for securing the sailstocks in the poll-end	38, 138
PORTER'S LOAD	Twenty stone (280 lb.—2½ cwt.)	111
POST-MILL	Mill with whole body revolving round a post	5, 17, 27, 63, 80, 114, 132
PRICK-POST	Central post in head of post-mill, under wind-shaft	23, 143
PRITCHELL	Pointed tool resembling a cold chisel, for dressing furrows of a millstone, and used with a hammer	49, 103
PUMPING-MILL	Marsh-mill with reciprocating pump instead of wheel	69
PURCHASE SHAFT	Shaft carrying purchase wheel for patent sail control	59
QUANT-BEARING	Glut-box (q.v.) supporting a quant post or spindle in an overdriven mill	48
QUARTER-BARS	The bars supporting main-post from the cross-trees	18, 83, 149
QUARTERN LOAF	Four pounds (the true quartern is 4 lb. 5½ oz.)	110
RABBET	Spring stick and cord holding feed shoe against damsel	53
RACK	Straight cog engaging with pinion; also a jack ring (q.v.)	44, 59

TRAM FRAME	Frame carrying the tram-wheels on the steps of a fantail post-mill. The wheels run on a tramway	64, 85, 94
TRANSOMS	Horizontal timbers in walls of smock-mills, etc.	23, 25
TREE-NAIL	Roughly made dowel-pin, also called a trunnel	74
TRIANGLES	Triangular links connecting rein irons with sail-rods in the 'patent' gear	59, 88
TROLLEY OR TRUCK WHEELS	Wheels carrying cap of smock- or tower-mill	29
TRUE OUT O'WIND	A dead true face to a length of timber	73
TRUNDLE WHEEL	Half-lantern pinion—pegs fitted into a disk	11
TRUNNEL	*See* Tree-nail	
TRUNNIONS	Bearings fitted in ends of horizontal block or spindle	35, 48, 98
TRYING-STICKS	Two sticks for checking a true face on timber	74
TURNBUCKLE	Double ended screw for drawing two rods together	22, 143
TUSK TENON	Tenon extended through a timber and pegged at tail end, as in the collar or 'girdle'	22
TWIBILL	Ancient tool for hacking out mortises	71
UNDERDRIFT MILL	With millstones driven from beneath the floor	44, 110
UPLONGS	The intermediate longitudinals in a common sail	39, 136
UPPER SIDE-GIRTS	Side-beams supporting the tail-beam under the wind-shaft	37
UPRIGHT SHAFT	The main driving shaft of the mill	43, 66, 121
VANES	Shutters in spring and patent sails, also the wings of the fantail	7, 10, 58, 64
VATS	In a corn-mill, the stone cases	50
V-PULLEY	Belt pulley having a v groove	63
WAIST OF MILL-STONE	The centre or 'eye' of the stone, where grain enters	49, 101
WALLOWER	Bevel pinion on upright shaft, engaging with brake-wheel	43, 66
WALL-PLATE	Wooden fillet in wall for fixing partitions	28
WARBLER	The bell-alarm (q.v.)	54, 126
WEATHER-BEAM	Main lateral beam beneath wind-shaft. It supports weather-studs secured by weather-irons	23, 30, 85, 143
'WEATHER' OF SAILS	The angularity or warp to catch the wind	7, 40, 6.
WEATHER SHUTTER	Trap-door over the canister	22, 65
WEATHER STUDS	Upright timbers either side of wind-shaft neck	23, 30
WHEATMEAL	Meal as it comes from the millstones (not dressed)	55, 108

BIBLIOGRAPHY

Batten, M. I.	*English Windmills* Vol 1 (SPAB 1930)
Bennett & Elton.	*History of Corn Milling* 4 vols (Simpkin Marshall 1898–1904)
Clarke, Allen.	*Windmill Land* 2 vols (Foulsham 1916–17)
Fairbairn, Sir William.	*Mills & Millwork* 2 vols (Longmans Green 1861)
Farries, K. G. & Mason, M. T.	*Windmills of Surrey & Inner London* (Skilton 1966)
Finch, Coles.	*Watermills & Windmills* (Daniel 1933)
Harrison, H. C.	*The Story of Sprowston Mill* (Phoenix House 1949)
Hemming, Rev P.	*Windmills in Sussex* (Daniel 1936)
Hopkins, Thurston.	*Old English Windmills & Inns* (Palmer 1927)
Hopkins, Thurston.	*Old Watermills & Windmills* (Alan 1933)
Hopkins, T. & Freese, S.	*In Search of English Windmills* (Palmer 1931)
Skilton, C. P.	*British Windmills & Watermills* (Collins 1948)
Smith, D.	*English Windmills* Vol 2 (SPAB 1932)
Vince, J. N. T.	*Discovering Windmills* (Shire 1969)
Vowles, Hugh P.	*The Quest for Power* (Chapman & Hall 1931)
Wailes, Rex.	*Windmills in England* (Architectural Press 1948)
Wailes, Rex.	*The English Windmill* (Routledge 1954)
Wolff.	*The Windmill as a Prime Mover* (Wiley, New York 1885)

Plate 1. Friston Mill, Saxmundham

Plate 2. Bledlow Ridge Mill, High Wycombe

Plate 3. Stevington Mill, Bedford

Plate 4. Cuddington Mill, Aylesbury

Plate 5. Monk Soham Oak Mill, Suffolk

Plate 6. Shipley Mill, Sussex

Plate 7. West Wratting Mill, Cambridgeshire

Plate 8. Cranbrook Mill, Kent

Plate 9. Old Marsh Mill, Tunstall Bridge, Norfolk

Plate 10. Polegate Mill, Sussex

Plate 11. *Heckington Mill, Lincolnshire*

(a)

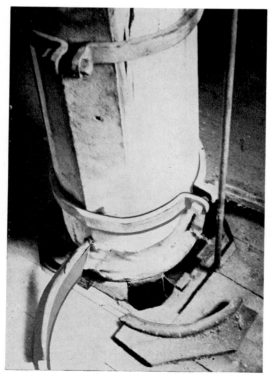

(b)

Plate 12. Substructure of a post-mill

(a)

(b)

(c)

(d)

Plate 13. The framing of windmills

Plate 14. Wind-shafts and bearings

(a)

(b)

Plate 15. Stone-floor and ground-floor

(a)

(b)

(c)

(d)

(e)

Plate 16. Wheels, millstones and governors

(a)

(b)

Plate 17. Bolter and dresser, Upminster Smock-Mill, Essex

Plate 18. Moving a main-post

(a)

(b)

Plate 19. Main-post and weather-beam

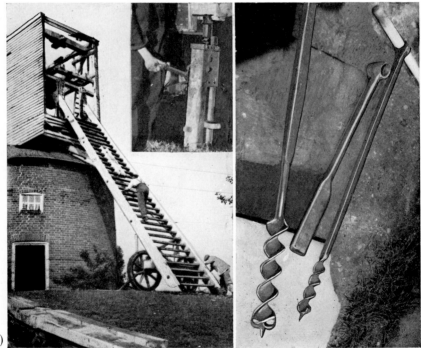

(a)

(b)

(c)

Plate 20. Tackle and tools

(a)   (b)

(c)

Plate 21.   *Mill carpentry*

(b)

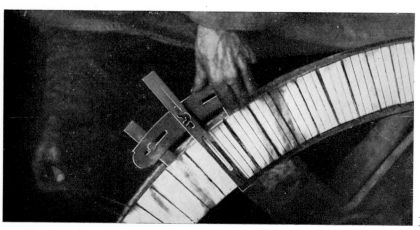

(a)

Plate 22. Cogging the mill

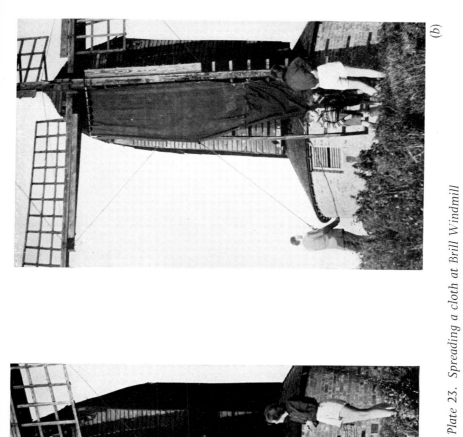

(b)

(a)

Plate 23. Spreading a cloth at Brill Windmill

(a)

(b)

(c)

Plate 24. *Spring sails*

Plate 25. Patent sails

Plate 26. Two old millers

(a)

(b)

Plate 27. Fantails and staging

(a)

(b)

(c)

(b)

(a)

Plate 28. Stone-dressing

(c)

(d)

(a)

(b)

Plate 29. *New sails at Brill*

Plate 30. A common-sailed paddle-wheel mill

(a)

(b)

Plate 31. Sail-frame and neck-brass

Plate 32. Worlingworth Mill, Suffolk

Plate 33. Cross-in-Hand Mill and Model, Sussex

Plate 34. Headcorn Mill, Kent

Plate 35. Coleby Heath six-sailer, Lincolnshire